PRINCIPLES AND MANAGEMENT OF CLOGGING IN MICRO IRRIGATION

Innovations and Challenges in Micro Irrigation

VOLUME 1

PRINCIPLES AND MANAGEMENT OF CLOGGING IN MICRO IRRIGATION

Edited by

Megh R. Goyal, PhD, PE, Vishal K. Chavan, and Vinod K. Tripathi

Apple Academic Press Inc. | Apple Academic Press Inc.
3333 Mistwell Crescent | 9 Spinnaker Way
Oakville, ON L6L 0A2 | Waretown, NJ 08758
Canada | USA

© 2016 by Apple Academic Press, Inc.

First issued in paperback 2021

Exclusive worldwide distribution by CRC Press, a member of Taylor & Francis Group

No claim to original U.S. Government works

ISBN-13: 978-1-77463-586-5 (pbk)
ISBN-13: 978-1-77188-277-4 (hbk)

Library and Archives Canada Cataloguing in Publication

Principles and management of clogging in micro irrigation / edited by Megh R. Goyal, PhD, PE, Vishal K. Chavan, and Vinod K. Tripathi.

(Innovations and challenges in micro irrigation; volume 1)
Includes bibliographical references and index.
Issued in print and electronic formats.
ISBN 978-1-77188-277-4 (hardcover).--ISBN 978-1-77188-278-1 (pdf)
1. Microirrigation. 2. Microirrigation--Maintenance and repair. 3. Microirrigation--Manage-ment. I. Goyal, Megh Raj, author, editor II. Chavan, Vishal K., author, editor III. Tripathi, Vinod K., author, editor IV. Series: Innovations and challenges in micro irrigation; v. 1

S619.T74P75 2015   631.5'87   C2015-905608-X   C2015-905609-8

CIP data on file with US Library of Congress

Apple Academic Press also publishes its books in a variety of electronic formats. Some content that appears in print may not be available in electronic format. For information about Apple Academic Press products, visit our website at **www.appleacademicpress.com** and the CRC Press website at **www.crcpress.com**

CONTENTS

LIST OF CONTRIBUTORS

S. R. Bhakar, PhD
Associate Professor, Department of Soil & Water Engineering, College of Technology and Agricultural Engineering, Udaipur, Rajasthan, India.

Vishal Keshavrao Chavan, PhD
Assistant Professor and Senior Research Fellow in SWE, Agriculture University, Akola, Maharashtra, India. Website: www.pdkv.ac.in; Email: vchavan2@gmail.com

Santosh K. Deshmukh, PhD
Chief Coordinator, Sustainability (Jalgaon Area, India Environmental Services), Jain Irrigation Systems Ltd, Jain Plastic Park, N.H. No. 6, Bambhori, Jalgaon 425001, Maharashtra, India. Tel: +91-257-2258011, Fax: +91-257-2258111. Website: www.jains.com, Email: jisl@jains.com.

Megh R. Goyal, PhD, PE
Retired Professor in Agricultural and Biomedical Engineering from General Engineering Department, University of Puerto Rico, Mayaguez Campus; and Senior Technical Editor-in-Chief in Agriculture Sciences and Biomedical Engineering, Apple Academic Press Inc., PO Box 86, Rincon, PR 00677, USA. Email: goyalmegh@gmail.com

A. M. Michael, PhD
Former Professor/Director, Water Technology Center, IARI; Ex-Vice-Chancellor, Kerala Agricultural University, Trichur, Kerala, India.

H. K. Mittal, PhD
Associate Professor, Department of Soil & Water Engineering, College of Technology and Agricultural Engineering, Udaipur, Rajasthan, India

Miguel A. Muñoz-Muñoz, PhD
Ex-President of University of Puerto Rico, University of Puerto Rico, Mayaguez Campus, College of Agriculture Sciences, Call Box 9000, Mayagüez, PR 00681-9000, USA. Tel: +1-787-265-3871. Email: miguel.munoz3@upr.edu

M. B. Nagdeve, PhD
Email: nagdeve@lycos.com

V. G. Nimkale, M Tech
Department of Soil & Water Engineering, College of Technology and Agricultural Engineering, Udaipur, Rajasthan, India. Email: vinodnimkale@gmail.com

Felix B. Reinders, PhD
Agricultural Research Council, Institute for Agricultural Engineering, Pretoria, South Africa, Private Bag X519, Silverton 0127, South Africa. Email: reindersf@arc.agric.za

S. J., Supekar, PhD
Professor, Department of Irrigation and Drainage Engineering, College of Agricultural Engineering and Technology, Marathwada Krishi Vidyapeeth, Parbhani 431402, (MS) India.

Pawar Pramod Suresh, M Tech
Rural Development Officer, Indian Overseas Bank, Pohegaon, Dist. Ahmednagar, Maharashtra, India.
Tel: +91-9503759312. Email: talegaonbr@mumnsco.iobnet.co.in

Vinod Kumar Tripathi, PhD
Assistant Professor, Center for Water Engineering & Management, Central University of Jharkhand,
Ratu-Loharghat road, Brambe, Ranchi, Jharkhand, 835205-India. Tel: +918987661439. Email: tripathi-wtcer@gmail.com

B. Upadhyay, PhD
Associate Professor, Department of Soil & Water Engineering, College of Technology and Agricultural
Engineering, Udaipur, Rajasthan, India.

LIST OF ABBREVIATIONS

$\varnothing$w	dry weight basis
ASABE	American Society of Agricultural and Biological Engineers
DIS	drip irrigation system
DOY	day of the year
EPAN	pan evaporation
ET	evapotranspiration
ETc	crop evapotranspiration
FAO	Food and Agricultural Organization, Rome
FC	field capacity
FUE	fertilizers use efficiency
gpm	gallons per minute
ISAE	Indian Society of Agricultural Engineers
kc	crop coefficient
kg	kilograms
Kp	pan coefficient
lps	liters per second
lph	liter per hour
msl	mean sea level
PE	polyethylene
PET	potential evapotranspiration
pH	acidity/alkalinity measurement scale
PM	Penman–Monteith
ppm	one part per million
psi	pounds per square inch
PVC	polyvinyl chloride
PWP	permanent wilting point
RA	extraterrestrial radiation
RH	relative humidity
RMAX	maximum relative humidity
RMIN	minimum relative humidity
RMSE	root mean squared error
RS	solar radiation
SAR	sodium absorption rate
SDI	subsurface drip irrigation
SW	saline water

SWB	soil water balance
TE	transpiration efficiency
TEW	total evaporable water
TMAX	maximum temperature
TMIN	minimum temperature
TR	temperature range
TSS	total soluble solids
TUE	transpiration use efficiency
UC	uniformity coefficient
USDA	U.S. Department of Agriculture
USDA-SCS	U.S. Department of Agriculture-Soil Conservation Service
WSEE	weighed standard error of estimate
WUE	water use efficiency

LIST OF SYMBOLS

A	cross-sectional flow area (L^2)
AW	available water ($\ominus_w$, %)
Cp	specific heat capacity of air (J/(g·°C))
CV	coefficient of variation
D	accumulative intake rate (mm/min)
d	depth of effective root zone
D	depth of irrigation water (mm)
Δ	slope of the vapor pressure curve (kPa°C^{-1})
e	vapor pressure, (kPa)
e_a	actual vapor pressure (kPa)
E	evapotranspiration rate (g/(m^2 s))
Ecp	cumulative class A pan evaporation for 2 consecutive days (mm)
eff	irrigation system efficiency
E_i	irrigation efficiency of drip system
E_p	pan evaporation as measured by Class-A pan evaporimeter (mm/day)
Es	saturation vapor pressure (kPa)
E_{pan}	class A pan evaporation
ER	cumulative effective rainfall for corresponding 2 days (mm)
e_s	saturation vapor pressure (kPa)
$e_s - e_a$	vapor pressure deficit (kPa)
ET	evapotranspiration rate (mm/year)
ETa	reference ET, in the same water evaporation units as Ra
ETc	crop-evapotranspiration (mm/day)
ET_o	the reference evapotranspiration obtained using the Penman–Monteith method (mm/day)
ET_{pan}	the pan evaporation-derived evapotranspiration
EU	emission uniformity
F	flow rate of the system (GPM)
FC	field capacity (v/v, %)
G	soil heat flux at land surface (W/m^2)
H	the plant canopy height (m)

h	the soil water pressure head (L)
I	the infiltration rate at time t (mm/min)
IR	injection rate, GPH
IRR	irrigation
K	the unsaturated hydraulic conductivity (LT^{-1})
K_c	crop coefficient
Kc	crop coefficient for bearing "Kinnow" plant
Kp	Pan factor
K_p	Pan coefficient
n	number of emitters
P	percentage of chlorine in the solution*
Pa	atmospheric pressure (Pa)
PWP	permanent wilting point (Θ_w%)
Q	flow rate (gpm)
q	the mean emitter discharges of each lateral (L h^{-1})
R	rainfall
r_a	aerodynamic resistance (s m^{-1})
Ra	extraterrestrial radiation, in the same water evaporation units as ETa
R_e	effective rainfall depth (mm)
R_i	individual rain gauge reading (mm)
R_n	net radiation at the crop surface (MJ m^{-1} day^{-1})
Rs	incoming solar radiation on land surface, in the same water evaporation units as ETa
RO	surface runoff
r_s	the bulk surface resistance (s m^{-1})
S	the sink term accounting for root water uptake (T^{-1})
Se	the effective saturation
S_p	plant-to-plant spacing (m)
S_r	row-to-row spacing (m)
SU	statistical uniformity (%)
S_ψ	water stress integral (MPa day)
t	the time that water is on the surface of the soil (min)
T	time (h)
V	volume of water required (liter/day/plant)
V_{id}	irrigation volume applied in each irrigation (L tree^{-1})
V_{pc}	the plant canopy volume (m^3)
W	the canopy width
W_p	fractional wetted area
z	the vertical coordinate positive downward (L)

α it is related to the inverse of a characteristic pore radius (L^{-1})

γ the psychrometric constant ($kPa\,°C^{-1}$)

θ volumetric soil water content (L^3L^{-3})

$\theta(h)$ the soil water retention (L^3L^{-3})

θr the residual water content (L^3L^{-3})

θ_s the saturated water content (L^3L^{-3})

θ_{vol} a volumetric moisture content (cm^3/cm^3)

λ latent heat of vaporization ($MJ\,kg^{-1}$)

λE latent heat flux (W/mo)

ρa mean air density at constant pressure ($kg\,m^{-3}$)

PREFACE BY MEGH R. GOYAL

Due to increased agricultural production, irrigated land has increased in the arid and subhumid zones around the world. Agriculture has started to compete for water use with industries, municipalities, and other sectors. This increasing demand along with increments in water and energy costs have made it necessary to develop new technologies for the adequate management of water. The intelligent use of water for crops requires understanding of evapotranspiration processes and use of efficient irrigation methods.

Every day, news on water scarcity appear throughout the world indicating that government agencies at central/state/local level, research and educational institutions, industry, sellers, and others are aware of the urgent need to adopt micro irrigation technology that can have an irrigation efficiency up to 90% compared with 30–40% for the conventional gravity irrigation systems. I stress the urgent need to implement micro irrigation systems in water scarcity regions.

Irrigation has been a central feature of agriculture since the start of civilization and the basis of the economy and society of numerous societies throughout the world. Among all irrigation systems, micro irrigation has the highest irrigation efficiency and is the most efficient. Micro irrigation is sustainable and is one of the best management practices. The water crisis is getting worse throughout the world, including Middle East and Puerto Rico where I live. We can therefore conclude that the problem of water scarcity is rampant globally, creating the urgent need for water conservation. The use of micro irrigation systems is expected to result in water savings and increased crop yields in terms of volume and quality. The other important benefits of using micro irrigation systems include expansion in the area under irrigation, water conservation, optimum use of fertilizers and chemicals through water, and decreased labor costs, among others. The worldwide population is increasing at a rapid rate, and it is imperative that food supply keeps pace with this increasing population.

Micro irrigation, also known as trickle irrigation or drip irrigation or localized irrigation or high-frequency or pressurized irrigation, is an irrigation method that saves water and fertilizer by allowing water to drip slowly to the roots of plants, either onto the soil surface or directly onto the root zone, through a network of valves, pipes, tubing, and emitters. It is done through narrow tubes that deliver water directly to the base of the plant. It allows controlled delivery of water directly to individual plants and can be installed on the soil surface or subsurface. Micro irrigation systems are often used for farms and large gardens, but are equally effective in the home garden or even for houseplants or lawns.

Water and fertigation management are important practices for the success of drip, micro, or trickle irrigation. Water management is the activity of planning, developing, distributing, and using water resources in an optimum way under defined water policies and regulations. It includes management of water treatment of drinking water or industrial water; sewage or wastewater; management of water resources in agriculture; management of flood protection; management of irrigation; management of the water table; and management of drainage.

A formidable obstacle to the successful operation of the system over its intended life of service is clogging of emitters. High water application uniformity is one of the significant advantages that a properly designed and maintained drip system can offer over other methods of irrigation. In many cases, the yield of crops may be directly related to the uniformity of water application. Partial or complete clogging drastically affects water application uniformity and, hence, may put a complete system out of operation, causing heavy loss to the crop and damage to the system itself. Thus, emitter clogging can nullify all the advantages of drip irrigation.

Since drip irrigation is expensive, its longevity must be maximized to assure a favorable benefit–cost ratio. If emitters get clogged in a short time after their installation, reclamation procedures to correct clogging increase maintenance cost and unfortunately may not be permanent. Clogging problems often discourage operators and consequently cause them to abandon the system and return to less efficient methods of irrigation.

Emitter clogging is directly related to the quality of irrigation water. Large quantities of irrigation water are obtained from underground sources. Since calcareous formations are considered to be good aquifers, water pumped from wells in these areas is generally rich in calcium carbonate and bicarbonate. Similar enrichment of water occurs as it comes in contact with other minerals, but dissolution and precipitation of calcium carbonates is most common. Precipitation of calcium carbonates may occur through the drip system, but the problem is most acute when it occurs within the narrow passages of the emitters or at the outlets. Fertilizers added to irrigation water greatly change the precipitation properties of the water if it is calcareous. Mineral precipitates (often seen as scale deposits), algae, and bacteria clog drip emitters. Clogged emitters result in variable distribution during irrigation and uneven fertilizer application during fertigation and thus hinder uniform crop development, reduce yield, and jeopardize quality. For growers to effectively use drip technology, they must prevent clogging of drip emitters. In this book, authors discuss innovations to alleviate problems of emitter clogging.

The mission of this compendium is to serve as a reference manual for graduate and undergraduate students of agricultural, biological, and civil engineering; horticulture; soil science; crop science; and agronomy. I hope that it will be a valuable reference for professionals that work with micro irrigation, wastewater, and water management, and for professional training institutes, technical agricultural centers, irrigation centers, agricultural extension service, and other agencies that work with micro irrigation programs.

After getting response from international readers for my first textbook on drip/ trickle or micro irrigation management by Apple Academic Press Inc., they have published for the world community this 10-volume series on *Research Advances in Sustainable Micro Irrigation*, edited by Megh R. Goyal. These 10 volumes are summarized at weblink appleacademicpress.com.

Volume 1 is the first volume of the series *Innovations and Challenges in Micro Irrigation*. This series is a must for those interested in irrigation planning and management, namely, researchers, scientists, educators, and students.

The contributions by the contributing authors have been most valuable in the compilation of this book. Their names are mentioned in each chapter and in the list of contributors. This book would not have been written without the valuable cooperation of these investigators, many of whom are renowned scientists who have worked in the field of micro irrigation throughout their professional careers.

I would like to thank editorial staff, Sandy Jones Sickels, Vice-President, and Ashish Kumar, Publisher and President at Apple Academic Press, Inc., for publishing this book when the diminishing water resources is a major issue worldwide. Special thanks are due to the AAP Production Staff for typesetting the entire manuscript and for the quality production of this book.

I express my deep admiration to my wife, Subhadra D Goyal, and my family for understanding and collaborating during the preparation of this book. With my whole heart and best affection, I dedicate this book to my father Late Shri Babu Ram Hira Lal, who was instrumental in shaping my life. My grandparents died when my father was 2 years old, and thereafter he was raised by his eldest sister. Later during his youth, he entered into a family business with his brother-in-law (Late Shri Mangoo Ram), who was among top three richest business men in the city of Sangrur (India). I remember that my father was treated as a slave by his master, and that led us to extreme poverty, eating only three meals a week (at least I). We were liberated from slavery when I was in seventh grade. Soon we bought an animal shelter house that was restored for human living with my carpentry skills. I inherited many qualities from my father: love of God and your family, neighbor, and country; do not harm anyone; never steal the property of others; always work with vocation and full speed without complaint; never expect any recognition; and work for others and the country. With greatest respect and honor, I salute to the humbleness, honesty, devotion, and vocation of my father. When my mind felt clogged with garbage, his patience was able to unclog it. I narrate a dream with him: Two days before his heavenly departure, he took me to all the religious places in Himalayas until we said good-bye (in a dream) to one another at the end of the earth. His memories have marked my heart, soul, and spirit.

As an educator, there is a piece of advice to one and all in the world: *Permit that our Almighty God, our Creator and excellent Teacher, irrigate the life with His Grace of rain trickle by trickle, because our life must continue trickling on... and never permit that our mind is clogged with futile things and garbage.*

— Megh R. Goyal, PhD, PE, Senior Editor-in-Chief
June 30, 2015

WARNING/DISCLAIMER

The goal of this compendium, **Principles and Management of Clogging in Micro Irrigation**, is to guide the world community on how to manage efficiently for economical crop production. The reader must be aware that dedication, commitment, honesty, and sincerity are the most important factors in a dynamic manner for complete success. This reference is not intended for a one-time reading; we advise you to consult it frequently. To err is human. However, we must do our best. Always, there is a place for learning from new experiences.

The editor, the contributing authors, the publisher, and the printer have made every effort to make this book as complete and as accurate as possible. However, there still may be grammatical errors or mistakes in the content or typography. Therefore, the contents in this book should be considered as a general guide and not a complete solution to address any specific situation in irrigation. For example, one size of irrigation pump does not fit all sizes of agricultural land and will not work for all crops.

The editor, the contributing authors, the publisher, and the printer shall have neither liability nor responsibility to any person, organization, or entity with respect to any loss or damage caused, or alleged to have caused, directly or indirectly, by information or advice contained in this book. Therefore, the purchaser/reader must assume full responsibility for the use of the book or the information therein.

The mention of commercial brands and trade names are only for technical purposes and does not imply endorsement. The editor, contributing authors, educational institutions, and the publisher do not have any preference for a particular product.

All weblinks that are mentioned in this book were active on June 30, 2015. The editors, the contributing authors, the publisher, and the printing company shall have neither liability nor responsibility if any of the weblinks are inactive at the time of reading of this book.

OTHER BOOKS ON MICRO IRRIGATION TECHNOLOGY FROM AAP

Management of Drip/Trickle or Micro Irrigation
Megh R. Goyal, PhD, PE, Senior Editor-in-Chief

Evapotranspiration: Principles and Applications for Water Management
Megh R. Goyal, PhD, PE, and Eric W. Harmsen, Editors

BOOK SERIES: RESEARCH ADVANCES IN SUSTAINABLE MICRO IRRIGATION
Senior Editor-in-Chief: Megh R. Goyal, PhD, PE

Volume 1: Sustainable Micro Irrigation: Principles and Practices
Senior Editor-in-Chief: Megh R. Goyal, PhD, PE

Volume 2: Sustainable Practices in Surface and Subsurface Micro Irrigation
Senior Editor-in-Chief: Megh R. Goyal, PhD, PE

Volume 3: Sustainable Micro Irrigation Management for Trees and Vines
Senior Editor-in-Chief: Megh R. Goyal, PhD, PE

Volume 4: Management, Performance, and Applications of Micro Irrigation
Senior Editor-in-Chief: Megh R. Goyal, PhD, PE

Volume 5: Applications of Furrow and Micro Irrigation in Arid and Semi-Arid Regions
Senior Editor-in-Chief: Megh R. Goyal, PhD, PE

Volume 6: Best Management Practices for Drip Irrigated Crops
Editors: Kamal Gurmit Singh, PhD, Megh R. Goyal, PhD, PE, and
Ramesh P. Rudra, PhD, PE

Volume 7: Closed Circuit Micro Irrigation Design: Theory and Applications
Senior Editor-in-Chief: Megh R. Goyal, PhD; Editor: Hani A. A. Mansour, PhD

Volume 8: Wastewater Management for Irrigation: Principles and Practices
Editor-in-Chief: Megh R. Goyal, PhD, PE; Coeditor: Vinod K. Tripathi, PhD

Volume 9: Water and Fertigation Management in Micro Irrigation
Senior Editor-in-Chief: Megh R. Goyal, PhD, PE

Volume 10: Innovations in Micro Irrigation Technology
Senior Editor-in-Chief: Megh R. Goyal, PhD, PE;
Coeditors: Vishal K. Chavan, MTech, and Vinod K. Tripathi, PhD

BOOK SERIES: INNOVATIONS AND CHALLENGES IN MICRO IRRIGATION

Senior Editor-in-Chief: Megh R. Goyal, PhD, PE

Volume 1: Principles and Management of Clogging in Micro Irrigation
Editors: Megh R. Goyal, PhD, PE, Vishal K. Chavan, and Vinod K. Tripathi

Volume 2: Sustainable Micro Irrigation Design Systems for Agricultural Crops: Methods and Practices
Editors: Megh R. Goyal, PhD, PE, and P. Panigrahi, PhD

Volume 3: Performance Evaluation of Micro Irrigation Management: Principles and Practices
Editor: Megh R. Goyal, PhD, PE,

Volume 4: Potential of Solar Energy and Emerging Technologies in Sustainable Micro Irrigation
Editors: Megh R. Goyal and Manoj K. Ghosal

ABOUT THE SERIES EDITOR-IN-CHIEF

Megh R. Goyal, PhD, PE

Megh R. Goyal, PhD, PE, is at present a retired professor in Agricultural and Biomedical Engineering from the General Engineering Department in the College of Engineering at the University of Puerto Rico, Mayaguez Campus, and Senior Acquisitions Editor and Senior Technical Editor-in-Chief in Agricultural and Biomedical Engineering for Apple Academic Press Inc. (AAP).

He received his BSc degree in engineering in 1971 from Punjab Agricultural University, Ludhiana, India; his MSc degree in 1977; his PhD degree in 1979 from the Ohio State University, Columbus; and his Master of Divinity degree in 2001 from Puerto Rico Evangelical Seminary, Hato Rey, Puerto Rico, USA.

Since 1971, he has worked as Soil Conservation Inspector (1971); Research Assistant at Haryana Agricultural University (1972–1975) and the Ohio State University (1975–1979); Research Agricultural Engineer/Professor at the Department of Agricultural Engineering of UPRM (1979–1997); and Professor in Agricultural and Biomedical Engineering at General Engineering Department of UPRM (1997–2012). He spent 1-year sabbatical leave in 2002–2003 at Biomedical Engineering Department, Florida International University, Miami, USA.

He was the first agricultural engineer to receive the professional license in Agricultural Engineering in 1986 from the College of Engineers and Surveyors of Puerto Rico. On September 16, 2005, he was proclaimed as "Father of Irrigation Engineering in Puerto Rico for the twentieth century" by the ASABE, Puerto Rico Section, for his pioneer work on micro irrigation, evapotranspiration, agroclimatology, and soil and water engineering. During his professional career of 45 years, he has received awards such as Scientist of the Year, Blue Ribbon Extension Award, Research Paper Award, Nolan Mitchell Young Extension Worker Award, Agricultural Engineer of the Year, Citations by Mayors of Juana Diaz and Ponce, Membership Grand Prize for ASAE Campaign, Felix Castro Rodriguez Academic Excellence, Rashtrya Ratan Award and Bharat Excellence Award and Gold Medal, Domingo Marrero Navarro Prize, Adopted Son of Moca, Irrigation Protagonist of UPRM, Man of Drip Irrigation by Mayor of Municipalities of Mayaguez/Caguas/Ponce, and Senate/Secretary of Agriculture of ELA, Puerto Rico.

He has authored more than 200 journal articles and textbooks, including Elements of Agroclimatology (Spanish) by UNISARC, Colombia, and two Bibliographies on Drip Irrigation. Apple Academic Press Inc. (AAP) has published his books, namely *Management of Drip/Trickle or Micro Irrigation, Evapotranspiration: Principles and Applications for Water Management,* and *Sustainable Micro Irrigation Design Systems for Agricultural Crops: Practices and Theory*, among others. He is the editor-in-chief of the book series Innovations and Challenges in Micro Irrigation and the 10-volume Research Advances in Sustainable Micro Irrigation. Readers may contact him at goyalmegh@gmail.com.

ABOUT THE COEDITORS

Vishal Keshavrao Chavan, MTech

Vishal Keshavrao Chavan, MTech, is currently working as a Senior Research Fellow in the Office of the Chief Scientist under Dr. Mahendra B. Nagdeve of the AICRP for Dryland Agriculture at Dr. Punjabrao Deshmukh Krishi Vidyapeeth, Akola, the premier agricultural university in Maharashtra, India. His work included the evaluation of the basic mechanism of clogging in drip irrigation system. His area of interest is primarily micro irrigation and soil and water engineering. He obtained his BTech degree in agricultural engineering in 2009 from Dr. Punjabrao Deshmukh Krishi Vidyapeeth, Akola, in Maharashtra, India; and his MTech degree from the University of Agricultural Sciences, in Raichur, Karnataka, India, in soil and water engineering. An expert on clogging mechanisms, he has published four popular articles and three research papers in national journals, and he attends international conferences. Readers may contact him at vchavan2@gmail.com

Vinod Kumar Tripathi, PhD

Vinod Kumar Tripathi, PhD, is working as Assistant Professor in the Center for Water Engineering & Management at the School of Natural Resources Management, Central University of Jharkhand, Brambe, Ranchi, Jharkhand, India.

He obtained his BTech degree in agricultural engineering in 1998 from Allahabad University, India, his MTech degree in irrigation and drainage engineering from G. B. Pant University of Agriculture & Technology, Pantnagar, India, and his PhD degree from Indian Agricultural Research Institute, New Delhi, India, in 2011.

He has developed a methodology to improve the quality of produce by utilizing municipal wastewater under drip irrigation during PhD research. His area of interest is geo-informatics, hydraulics in micro irrigation, and development of suitable water management technologies for higher crop and water productivity. He has taught hydraulics, design of hydraulic structures, water and wastewater engineering, and geo-informatics to the postgraduate students at Central University of Jharkhand, Ranchi, India.

He is an international expert and has evaluated funded schemes for rural water supply and sanitation. He is a critical reader, thinker, planner, and fluent writer and has published more than 32 peer-reviewed research publications and bulletins and has attended several national/international conferences. Readers may contact him attripathiwtcer@gmail.com

BOOK REVIEWS

This book series is informative and is a must for all irrigation planners to minimize the problem of water scarcity worldwide. Father of Irrigation Engineering in Puerto Rico of 21st Century and pioneer on micro irrigation in the Latin America, Dr. Goyal [my longtime colleague] has done an extraordinary job in the presentation of this book volume.

— Miguel A. Muñoz, PhD., Ex-President of University of Puerto Rico, and Professor/ Soil Scientist

I recall my association with Dr. Megh Raj Goyal while at Punjab Agricultural University, India. I congratulate him on his professional contributions and his distinction in micro irrigation. I believe that this innovative book series will aid the irrigation fraternity throughout the world.

— A. M. Michael, PhD, Former Professor/Director, Water Technology Center, IARI; Ex-Vice-Chancellor, Kerala Agricultural University, Trichur, Kerala, India

PART I

PRINCIPLES OF CLOGGING

CHAPTER 1

PRINCIPLES OF CHLORATION AND ACIDIFICATION

MEGH R. GOYAL

CONTENTS

This chapter is modified and printed with permission from "Goyal, M. R., Ed., 2013. *Management of Drip/Trickle or Micro Irrigation*. Chapters 8, 9, and 11, pp. 165–194 and 213–218. Oakville, ON, Canada: Apple Academic Press Inc., ISBN 9781926895123." ©2013.

1.1 INTRODUCTION

During recent years, the application of chemicals through drip irrigation was adopted. Thus, new technical terms such as chemigation, chloration, pestigation, insectigation, fungigation, nemagation, and herbigation were defined.[1–3] The term chemigation was adopted to include the application of chemicals through irrigation systems. Chemigation offers the following economic advantages compared with other conventional methods[3]:

1. Provides uniformity in the application of chemicals allowing the distribution of these in small quantities during the growing season when and where these are needed.
2. Reduces soil compaction and the chemical damage to the crop.
3. Reduces the quantity of chemicals used in the crop and the health hazards and risks during the application.
4. Reduces pollution of the environment.
5. Reduces the costs of manual labor, equipment, and energy.

If the chemical analysis shows a high concentration of salts, it can cause clogging problems.[4] Therefore, it is recommended to avoid the use of chemicals that can cause precipitates. For a better efficiency in application, the chemicals should be distributed uniformly around the plants. The uniformity of chemical distribution depends on:

1. the efficiency of the mixture,
2. the uniformity of the water application,
3. the characteristics of the flow,
4. the elements or chemical compounds that are present in the soil.

1.2 CHEMICAL INJECTION METHODS[5–7]

1.2.1 SELECTION OF A PUMP FOR THE INJECTION OF A CHEMICAL SUBSTANCE

While selecting an injection pump, one must consider the parts that will be in direct contact with the chemical substances. These parts must be of stainless steel or of a material that is corrosion resistant. The injection method consists of basic components such as pump, pressure regulator, gate valve, pressure gauge, connecting tubes, check valve, and chemigation tank. The injection pump should be precise, easy to adjust for different degrees of injection, corrosion resistant, durable, and rechargeable, with the availability of spare parts. The recommended materials are stainless steel, resistant plastic, rubber, and aluminum. Bronze, iron, and copper are nonacceptable materials. The injection pump must provide a pressure on the discharge line greater than the irrigation pump. Therefore, the operating pressure of the irrigation system exerts a minimum effect in the endurance of the injection pump.

1.2.2   INJECTION METHODS

The efficiency of chemigation depends on the capacity of the injection tank, solubility of a chemical in water, dilution ratio, precision of dilution, potability, costs and the capacity of the unit, method of operation, experience of the operator, and needs of the operator. The chemical compounds used for the chemigation process must be liquid emulsions or soluble powder. The injector must be appropriate to introduce substances in the system. In addition, it must be of an adequate size to supply the necessary amount of chemical at a desired flow rate. Normally, the injection is carried out with an auxiliary electric pump or by interconnecting the injection pump to the irrigation pump.

1.2.2.1   INJECTION BY A PRESSURE PUMP

A rotary, diaphragm, or piston-type pump can be used to inject the chemicals to flow from the chemical tank toward the irrigation line. The chemigation pump must develop a pressure greater than the operating pressure in the irrigation line. The internal parts of the pump must be corrosion resistant. This method is very precise and reliable for injecting the chemicals in the drip irrigation system.

1.2.2.2   INJECTION BY PRESSURE DIFFERENCE

This is one of the easiest methods to operate. In this method, a low-pressure tank is used. This tank is connected with the discharge line at two points: one that serves as a water entrance to the tank and the other is an exit for the mixture of chemicals. A pressure difference is created with a gate valve in the main line. The pressure difference is enough to cause a flow of water through the tank. The chemical mixture flows into the irrigation line. The concentration of the chemical in water is difficult to calculate and control. Therefore, it is recommended to install an accurate metering valve to maintain a precalibrated injection rate.

1.2.2.3   INJECTION BY VENTURI PRINCIPLE

A Venturi system can be used to inject chemicals into the irrigation line. There is a decrease in the pressure accompanied by an increase in the liquid velocity through a Venturi. The pressure difference is created across the Venturi, and it is sufficient to cause a flow by the suction of chemical solutions from the tank.

1.2.2.4 INJECTION IN THE SUCTION LINE OF THE IRRIGATION PUMP

A hose or tube can be connected to the suction pipe of the irrigation pump to inject the chemicals. A second hose or tube is connected to the discharge line of the pump to supply water to the tank. This method should not be used with toxic chemical compounds because of the possible contamination of the water source. A foot valve or safety valve at the end of a suction line can avoid the contamination.

1.3 FERTIGATION

All fertilizers used for the chemigation purpose must be soluble (Table 1.1). The partially soluble chemical compounds can cause clogging and thus can create operational problems.[1,5]

TABLE 1.1 Solubility of Commercial Fertilizers

Fertilizer	Solubility (g/L)
Ammonia	97
Ammonium nitrate	1185
Ammonium sulfate	700
Calcium nitrate	2670
Calcium sulfate	Insoluble
Diammonium phosphate	413
Dicalcium phosphate	Insoluble
Magnesium sulfate	700
Manganese sulfate	517
Monoammonium phosphate	225
Monocalcium phosphate	Insoluble
Potassium chloride	277
Potassium nitrate	135
Potassium sulfate	67
Urea	1190

1.3.1 NITROGEN

Nitrogen is an element that is most frequently applied in the drip irrigation system. The principal sources of nitrogen for chemigation are anhydrous ammonia, liquid ammonia, ammonium sulfate, urea, ammonium nitrate, and calcium nitrate. The anhydrous ammonia or the liquid ammonia can increase the pH of the irrigation water, thus a possible precipitation of calcium and magnesium salts. If the irrigation water has a high concentration of calcium and magnesium bicarbonates, then these can result in the precipitation of chemical compounds. This enhances the clogging problems in the drippers, filters, and laterals. The ammonium salts are very soluble in water and cause less problems of clogging, with the exception of ammonium phosphate. The phosphate salts tend to precipitate in the form of calcium and magnesium phosphates if there is an abundance of Ca and Mg in the irrigation water. Ammonium sulfate causes little obstruction problems or changes in pH water. The urea is very soluble, and it does not react with the irrigation water to form ions, unless water contains the enzyme urease. This enzyme can be present if water has large amounts of algae or other biological agents.

The filtration system does not remove urease. This can cause hydrolysis of urea. Because the concentrations of the enzyme are generally low compared with those in the soil, urea will not hydrolyze to a significant degree in the irrigation water. The nitrate salts (e.g., calcium nitrate) are relatively soluble in water and do not cause large changes in the pH of the irrigation water. The nitrogen fertigation is more effective than the conventional methods of application, especially in sandy soils. In addition, the nitrogen fertigation is more efficient than the conventional methods in fine-textured soils.

1.3.2 PHOSPHORUS

Phosphorus can be applied in the irrigation system as an organic phosphate compound and glycerophosphates. The organic phosphates (orthophosphates) and urea phosphate are relatively soluble in water and can easily move in the soil. The organic phosphates do not precipitate, and the hydrolysis of an organic phosphate requires large lapse time.

Glycerophosphates react with calcium to form compounds of moderate solubility. The application of phosphorus can enhance obstructions in the drip irrigation system. When phosphoric fertilizers are applied in the irrigation water with high concentrations of Ca and Mg, then the insoluble phosphate compounds are formed that can obstruct the drippers and lateral lines. Phosphorus moves slowly in the soil and the root zone. In addition, the moist soil particles absorb phosphorus to form insoluble compounds. It is not recommended to fertigate phosphorus fertilizers during the growth period of a crop. Instead, it should be applied before seeding, during seeding, and during fruit formation. The plant uses phosphorus early in its growth. In the drip-irrigated crops, the fertigation of phosphorus can be combined with traditional methods of application. The drip irrigation system is efficient in the application of soluble phosphorus compounds, because water is applied in the root zone, which facilitates the availability of phosphorus.

1.3.3 POTASSIUM

Potassium can be fertigated in the form of potassium sulfate, potassium chloride, and potassium nitrate. Generally, potassium salts have good solubility in water and cause little problems of precipitation.

1.3.4 MICRONUTRIENTS

The micronutrients are supplied in the form of chelates. Thus, its solubility in water is increased, and these do not cause any problems of obstruction and precipitation. If the micronutrients are not applied as recommended, then iron, zinc, copper, and magnesium can react with the salts in the soil causing precipitation. This enhances the clogging of the emitters. The chelates should be dissolved before fertigation. The chemigation of micronutrients benefits the plant to accomplish a good development and growth. In addition, the operational cost is lower compared with the foliage application.

1.4 CHLORATION OR CHLORINATION

The obstruction of the filters, the distribution lines, or the emitters is the main problem associated with the operation and management of a drip irrigation system.[8-10] These obstructions are caused by physical agents (solid particles in suspension), chemical agents (precipitation of insoluble compounds), or biological agents (macro- and microorganisms). The preventive maintenance is the best solution to reduce or to eliminate the obstructions in the emitters or components of the system.[11] The chlorination is an addition of chlorine to water.[10] Chlorine, when dissolves, acts as an oxidation agent and attacks the microorganisms, such as the algae, fungi, and bacteria. This procedure has been used for many decades to purify the drinking water.[12] Chloration is an injection of chlorine compounds through the irrigation system. The chloration solves the problem of obstruction of the emitters or drippers caused by biological agents effectively and economically. "Caution: Do not use any chemical agent through the drip system without consulting a specialist."[9,10,13,14]

The chloration is the cheapest and effective treatment for the control of bacteria, algae, and the slime in the irrigation water. Chlorine can be introduced at low concentrations (1 ppm) at necessary intervals, or at high concentrations (10–20 ppm) for a few minutes. Chlorine can be injected in form of chlorine powder (solid) and chlorine gas. The gas treatment is expensive and dangerous for the operator. Calcium hypochlorite can also be used, but calcium tends to precipitate. Chlorine also acts as a biocide to iron and sulfur bacteria.

1.4.1 QUALITY OF WATER

1.4.1.1 WATER SOURCE

It is necessary to conduct the physical and chemical analysis of water before designing a drip irrigation system and choosing an appropriate filtration system.[14] For the chemical analysis, it is important to take a representative sample of the water. If the water source is subsurface (e.g., deep well), the sample must be taken an hour and a half after the pump begins to work. When the water source is from a lake, river, pool, or open channel, samples must be taken at the surface, at the center, and at the bottom of the water source.

It is important to analyze the sample for suspended solids, dissolved solids and acidity (pH), macroorganisms, and microorganisms.[15] The acidity of the water must be known, because it is a factor that affects the "chemigation directly" and therefore the chloration. For example, the chloration for the control of bacteria is ineffective for a pH > 7.5. Therefore, it is necessary to add acid to lower the pH of irrigation water and to optimize the biocide action of the chlorine compound. If the chemical analysis of the water is known, we can predict the obstruction problems and take suitable measures. In addition, a program of adequate service and maintenance can be developed. The physical, chemical, and biological agents are classified in Table 1.2. The factors are classified in order of the risk: from low to severe. When the amount of solids, salts, and bacteria in the water is within the acceptable limits, the risk of clogging is reduced.

TABLE 1.2 Water Quality: Criterion that Indicates the Risk of Obstruction of the Emitters

Type of Problem	Risk of Obstruction		
	Low	Moderate	Severe
Biological agents			
Bacteria populationa	<10 000	10 000–50 000	>50 000
Physical agents			
Suspended solidsb	<50	50–100	>100
Chemical agents			
Acidity (pH)	<7.0	7.0–8.0	>8.0
Dissolved solidsb	<0.2	0.2–1.5	>2000

TABLE 1.2 *(Continued)*

Type of Problem	Risk of Obstruction		
	Low	Moderate	Severe
Ironb	<0.1	0.1–1.5	>1.5
Hydrogen sulfide	<500	500–2000	>2.0
Manganeseb	<0.2	0.2–2.0	>1.5

[a]Maximum concentration of the representative sample of water. Given in ppm (mg/L).
[b]Maximum number of bacteria per milliliter. Obtained from field samples and laboratory analysis.

Also the particles of "organic matter" can combine with bacteria and produce a type of obstruction that cannot be controlled with the filtration system. The fine particles of organic matter are deposited within the emitters and are cemented with bacteria such as *Pseudomonas* and *Enterobacter*. This combined mass causes clogging of the emitters. This problem can be controlled with super chloration at a rate of 1000 ppm (mg/L). However, the chloration at these high rates can cause toxicity of a crop. The obstruction caused by the "biological agents" constitutes a serious problem in the drip irrigation system that contains organic sediments with iron or hydrogen sulfide. Generally, the obstruction is not a serious problem if water does not have organic carbon, which is a power source for the bacteria (promotes the bacterial growth). There are several organisms that increase the probability of obstruction when there are ions of iron (Fe^{++}) or sulfur (S^-).

Algae in surface water can add carbon to the system. The slime can grow on the inner surface of the pipes. The combination of the fertilizer and the heating of the polyethylene pipes (black) because of sunlight can promote the formation and development of these microorganisms. Many of the water sources contain carbonates and bicarbonates, which serve like an inorganic power source to promote slime growth; also, autotrophic bacteria (that synthesize their own food) are developed. The algae and the fungi are developed in the surface waters. Besides obstructing the emitters, the filamentous algae form a gelatinous substance in the pipes and emitters, which serve as a base for the development of slime. Another type of obstruction can also happen when the filamentous bacteria precipitate iron into the insoluble iron compounds (Fe^{+++}).

1.4.1.2 GROWTH OF SLIME IN THE DRIP IRRIGATION SYSTEM

The bacteria can grow within the system in the absence of light and produce a mass of the slime or cause the precipitation of iron or sulfur dissolved in water. The slime can act like an adhesive substance that agglutinates fine clay particles sufficiently large enough to cause clogging.[8,10,13]

1.4.2   GROWTH OF ALGAE IN THE WATER SOURCE OR IN THE IRRIGATION SYSTEM

One of the most frequent problems is the growth of algae and other aquatic plants in the surface water that can be used for drip irrigation. The algae grow well in the surface water. The problem becomes serious if the water source contains nitrogen, phosphorus, or both. In many cases, the algae can cause obstructions in the filtration system. When the screen filters are used, the algae can be entangled in the sieves (screen) of the filter. In high concentration, these aquatic microorganisms can create problems in the sand filters. This requires a frequent flushing and cleaning of the filters.

1.4.3   PRINCIPLE OF CHLORATION[8,12,15–19]

The principle of chloration for treating water by drip irrigation is similar to the one that is used to purify water for drinking purpose. Table 1.3 includes basic reactions of chlorine and its salts. When chlorine in gaseous state (Cl_2) dissolves in water, the chlorine molecule is combined with water in a reaction called hydrolysis. The hydrolysis produces hypochloric acid (HOCl; reaction (1)). Following this reaction, hypochloric acid enters an ionization reaction as shown in reaction (2). Hypochloric acid (HOCl) and hypochlorite (OCl) are known as freely available compounds and are responsible for controlling the microorganisms in water. The equilibrium of these depends on the temperature and the pH of the irrigation water. When the water is acidic (low pH), the equilibrium moves to the left, resulting in an increase in HOCl. When the water is alkaline (high pH), chlorine increases in the form of OCl⁻. The efficiency of HOCl is 40–80 times greater than OCl⁻. Therefore, the efficiency of the chloration depends greatly on the acidity (pH) of the water source. Reaction (1) produces hydrogen ions (H^+) that can increase the acidity. The basicity depends on the amount of added chlorine and the buffer capacity of water. Sodium hypochlorite [NaOCl] and calcium hypochlorite [$Ca(OCl)_2$] hydrolyze and produce OH⁻ ions that tend to lower the acidity of water (reactions (3) and (4)). If the pH is extremely low, the gaseous chlorine (Cl_2) predominates, which can be dangerous. Therefore, it is recommended to store the sources of OCl compounds separate from solids. Also the available free chlorine reacts with oxidizing compounds (such as iron, manganese, and hydrogen sulfide) and produces insoluble compounds, which must be removed from the system to avoid clogging.

TABLE 1.3   Basic Forms of Chlorine Reactions and Its Salts

Reactions	Reaction Number
$Cl_2 + H_2O = H^+ + Cl^- + Col$	(1)
$HOCl = H^+ + OCl^-$	(2)
$NaOCl + H_2O = Na^+ + OH^- + Col$	(3)
$Ca(OCl)_2 + 2H_2O = Ca^{2+} + 2OH^- + 2HOCl$	(4)
$HOCl + NH_3 = NH_2Cl + H_2O$	(5)
$HOCl + NH_2Cl = NHCl_2 + H_2O$	(6)
$HOCl + NHCl_2 = NCl_3 + H_2O$	(7)
$HOCl + 2Fe^{2+} + H^+ = 2Fe^{3+} + Cl^- + H_2O$ Ferrous to ferric	(8)
$Cl_2 + 2Fe(HCO_3)_2 + Ca(HCO_3)_2 = 2Fe(OH)_3 \text{ (insoluble)} + CaCl_2 + 6CO_2$	(9)
$HOCl + H_2S = S—\text{(insoluble)} + H_2O + H^+ + Cl^-$	(10)
$Cl_2 + H_2S = S—\text{(insoluble)} + 2H^+ + 2Cl^-$	(11)

Chlorine has two important chemical properties: at a low concentration (1–5 mg/L), it acts as a bactericidal, and at a high concentration (100–1000 mg/L), it acts as an oxidizing agent, which can disintegrate particles of organic matter. It is necessary to watch, because chlorine at these high levels can affect the growth of some plants.

1.4.4   SOURCES OF COMMERCIAL CHLORINE

The most common chorine sources[17] used in a drip irrigation system are sodium hypochlorite, calcium hypochlorite, and gaseous chlorine.

1.4.4.1   SODIUM HYPOCHLORITE, NAOCL

Sodium hypochlorite is a liquid and is commonly used as whitener for clothes. It can be easily decomposed at high concentrations, in the presence of light and heat. It must be stored at room temperature in packages resistant to corrosion. This compound is easy to handle. The amounts can be measured precisely, and it causes few problems.

1.4.4.2 CALCIUM HYPOCHLORITE, CA(OCL)$_2$

Calcium hypochlorite is available commercially as dust, granulated, or pellets. It is well soluble in water and is quite stable under appropriate storage conditions. It must be stored at room temperature in a dry place and in packages resistant to corrosion. When this compound is mixed in a concentrated solution, it forms a suspension that contains calcium oxalate, calcium carbonate, and calcium hydroxide. These compounds can obstruct the drip irrigation system.

1.4.4.3 GASEOUS CHLORINE, CL$_2$ GAS

It is available in liquid form at high pressure in cylinders from 45 to 1000 kg. The Cl$_2$ is very poisonous and corrosive. It must be stored in a well-ventilated place. Table 1.4 shows equivalent amounts of chlorine for different commercial sources and the required amount to treat 1233 m³ (1 acre-foot) of water to obtain 1 ppm of chlorine. The NaOCl is safer than Cl$_2$ and avoids calcium precipitates in the emitters, which can happen when Ca(OCl)$_2$ is used. It is more economical to use than Cl$_2$ in large systems. In small systems, it is appropriate to use sodium or calcium hypochlorite. The use of Cl$_2$ is preferred in situations where the addition of sodium and calcium can be detrimental to the crop. It is necessary to observe that Cl$_2$ is dangerous under certain conditions. Thus, the instructions on the label must be followed. It is recommended to install a security valve (one way or check valve) in the tank that is used for injecting chlorine.

TABLE 1.4 Equivalent Amounts of the Commercial Sources of Chlorine and the Required Amount to Treat 1 acre-foot of Water to Obtain 1 ppm of Chlorine

Commercial Source of Chlorine	Equivalent Amount to Obtain 454 g (1 lb) of Chlorine	Required Amount to Treat 1 acre-foot (1233 m3) of Water and Obtain 1 ppm of Chlorine
Gaseous chlorine (Cl2)	454 g (1.0 lb)	1226 g (2.7 lb)
Calcium hypochlorite, Ca(OCl)2		
65–70% of available chlorine	681 g (1.5 lb)	1816 g (4.0 lb)
Sodium hypochlorite, NaOCl		
15% of available chlorine	2.54 L (0.67 gallons)	6.81 L (1.8 gallons)
10% of available chlorine	3.78 L (1.0 gallons)	10.22 L (2.7 gallons)
0.5% of available chlorine	7.57 L (2.0 gallons)	20.44 L (3.4 gallons)

1.4.5   METHOD OF CHLORATION[15,18]

The chloration in a drip irrigation system can be continuous or in intervals, depending on the desired results. Application at intervals is appropriate, when the objective is to control the growth of microorganisms in lateral lines, emitters, or other parts of the system. The continuous treatment is used when we want to precipitate the iron dissolved in water, to control algae in the system, or where it is not reliable to use the treatment at intervals.

1.4.5.1   GENERAL RECOMMENDATIONS

1. Inject chlorine before the filters. This controls the growth of algae or bacteria in the filters, which otherwise would reduce the filtration efficiency. This also allows the filtration of any precipitate caused by the injection of chlorine.
2. Calculate the amount of chlorine to be injected. It is necessary to know the volume of water to be treated, the active ingredient of the chemical compound to be used, and the desired concentration in the treated water.
3. Chlorine should be injected when the system is in operation.
4. One should take samples from water at the nearest drippers and the most distant drippers to determine the chlorine level at these points. Allow sufficient time so that the lines are filled with the chlorine solution.
5. Adjust the injection ratio. Repeat steps 4 and 5 until the desired concentration is obtained in the system.

1.4.5.2   RECOMMENDED CHLORINE CONCENTRATIONS[17]

1. Continuous treatment (in order to prevent the growth of algae or bacteria): Apply from 1 to 2 mg/L continuously through the system.
2. Treatment at intervals (in order to eliminate the algae or bacteria): Apply from 10 to 20 mg/L for 60 min. The frequency of the treatment depends on the concentration of these microorganisms in the water source.
3. Super chloration (in order to dissolve the organic matter and, in many cases, the calcium precipitated in the drippers): Inject chlorine at a concentration from 50 to 100 mg/L, depending on the case. After this, close the system and leave it for 24 h, to clean all the secondary and lateral lines. It helps to clean the obstructions in the secondary and lateral lines. We have to be careful while applying these amounts, because these chlorine levels can be toxic to certain crops. Table 1.5 shows typical dosages of chlorine.

TABLE 1.5   Typical Dosages of Chlorine

Problem due to	Dosage
Algae	1–2 ppm continuous, or 10–20 ppm for 30–60 min
Ferro bacteria	1 + ppm: varies with the amount of bacteria
Slime	0.5 ppm
Precipitation of iron	0.64 × [content of Fe++]
Precipitation of manganese	1.3 × [content of manganese]
Hydrogen sulfide	3.6 × 8.4 × [content of H2S]

1.4.5.3   CHLORINE REQUIREMENTS[12,15–18]

Chlorine requirements must be known before the chloration. The Cl_2 or NaOCl is a biocide that must be applied in the amounts and at recommended concentrations. The excess of chlorine in the irrigation water can cause damage to the young plants or young trees. On the other hand, the low levels do not solve the problems associated with the growth of microorganisms in the irrigation water. Chlorine is a very active and toxic agent at high concentrations; therefore, it must be handled carefully. When it is injected in the irrigation lines, some chlorine reacts with inorganic compounds and organic substances of the water or it adheres to them. In most wells and water sources, from 65 to 81% of chlorine is lost by this type of reaction. Chlorine (like hypochlorous acid) that adheres to the organic matter or that reacts with other compounds does not destroy microorganisms. For this reason, it does not have value as a biocide agent. The free chlorine (the excess of hypochlorous acid) is the agent that inhibits the growth of bacteria, algae, and other microorganisms in the water. Therefore, it is indispensable to establish the chlorine requirements before the chloration. In this way, we can maintain the desired concentrations of available chlorine.

In order to inhibit the growth of microorganisms, a minimum contact time of 30 min is required (45 min of injection). It also requires a minimum concentration of 0.5–1.0 mg/L of available chlorine measured at the end of the drip line and 2.0–3.0 mg/L of available chlorine at the injection point. The following equations are used to calculate the gallons per hour (gph) of NaOCl that must be injected to obtain the desired concentration of chlorine per minute (gpm):

1.   Formula for gpm of 10% NaOCl:

$$= [0.0006 \times (\text{gpm desirable chlorine}) \times (\text{discharge of the pump, gpm})] \qquad (1)$$

2.   Formula for gph of 5.25% NaOCl:

$$= [0.000114 \times (\text{ppm desirable chlorine}) \times (\text{discharge of the pump, gpm})] \tag{2}$$

3. Formula for pounds by hectare of Cl_2 (gas):

$= [0.06 \times \text{ppm of desirable chlorine} \times \text{discharge of the pump in gpm}] / [\text{percentage of chlorine in the material}]$ (3)

4. Gallons of liquid chlorine per hour:
$= [0.06 \times \text{ppm of desirable chlorine} \times \text{discharge of the pump in gpm}] / [\text{percentage of chlorine in the material}]$
(4)

5. Dry chlorine in pounds per hour:

$$= [0.05 \times \text{ppm} \times \text{gpm}] / [\text{percentage of chlorine in the material}] \tag{5}$$

6. Dry chlorine in pounds per 1000 gallons of water:

$$= [0.83 \times \text{ppm}] / [\text{percentage of chlorine in the material}] \tag{6}$$

7. For chlorine gas:

$= \text{Take the percentage of chlorine} = \text{as } 100; \text{and calculate as dry chlorine}$ (7)

1.4.5.4   HOW TO MEASURE CHLORINE WITH THE DPD METHOD

It is essential to measure chlorine when using liquid chlorine as bactericide and algicide in irrigation systems of low volume.[15,19] Most of the methods of measuring chlorine which are used in the swimming pools are not adequate for irrigation systems. This is because many of these equipments measure only the total chlorine, but not the residual free chlorine. The equipment using N,N-diethyl-p-phenylenediamine (DPD) of good quality can measure the total chlorine and the free available chlorine. The test equipment with DPD is very simple. The directions and procedures come with the equipment. The equipment is used to measure each type of chlorine. When applying these compounds, water becomes pink in the presence of chlorine. The more intense is the color, higher is the chlorine concentration. In order to know the chlorine concentration, the color of water is compared with that of a calibrated chromatic chart. One must remember that the free chlorine is the one that determines the biocide action. If free chlorine is not sufficiently available, the bacteria continue to grow even though chlorine has been injected into the system. In other words, if the amount of total chlorine is not sufficient to maintain free chlorine in solution, the treatment gets of no value. The test equipment with DPD can be purchased from the sellers of irrigation equipments or from chemical agents who are specialized in water treatment.

1.5 INSTALLATION, OPERATION, AND MAINTENANCE

1.5.1 INSTALLATION

To be effective, all types of chemigation equipments must be suitable, reliable, and precise. All the electrical devices must resist risks due to bad weather. In addition, all the valves, accessories, fittings, and pipes must resist the operating pressure. Some chemicals can cause corrosion problems. The chemical compounds and the concentrations of chemicals must be compatible with the injection system. The materials of the system must be corrosion resistant. These should be washed with clean water after each fumigation process.

1.5.2 OPERATION

The chemigation procedure must follow a preestablished order. Irrigation system is operated until the soil saturates to a field capacity. Then, the chemicals are injected. Once the chemigation has finished, water is allowed to flow free in the system for a sufficient time to remove all the sediments (salts) of chemicals from the laterals and drippers.

1.5.3 MAINTENANCE

The maintenance is a routine procedure. One should inspect all the components of the injection system after each application. It is recommended to replace the defective components before they stop working altogether. In order to clean the injection system, it is necessary to have water accessible near the system. After each application, it is recommended to clean and wash exterior of all the parts with water and detergent and to rinse with clean water. The chemigation system can be cleaned in two ways:
1. By pressurized air
2. By using acids or other chemical agents

The pressurized air is used to clean the laterals of accumulation of the organic matter. Also, the lines can be cleaned with a commercial grade hydrochloric acid, phosphoric acid, or sulfuric acid at a concentration of 33–38%. When the acid is used, it is important to use protective clothing to avoid risks and accidents due to burns. Before using acid, it is recommended to allow the water flow through the system for 15 min. It is safe to fill the tank to two-thirds of its capacity and make sure that all the components are in good condition. Now add the acid to the tank. The system will operate at a pressure of 0.8–1.0 atmospheres to apply acid. The acid treatment will avoid the precipitation of salts and the formation of slime in the system. When the treatment has been completed, allow the water to flow to remove the residues of acid. Other practices for a good operation are:

1. to lubricate the movable screws and parts, after using the system,
2. to lubricate the movable screws and parts if the fertigation system was inactive during a prolonged period. It is necessary to activate the system and to make sure that all the parts are in good condition.

"Rule of thumb is to chemigate during the middle of the irrigation cycle."

1.5.4   CALIBRATION

The calibration consists of adjustment of the injection equipment to supply a desired amount of chemicals. The adjustment is necessary to make sure that the recommended dosage is applied. Excess of chemicals is very dangerous and hazardous, whereas a small amount will not give effective results. The use of excess fertilizers is also not economical. The amounts less than the recommended dosages cause reduction in the crop yield. The precise calibration helps us to obtain accurate, reliable, and desirable results. A simple calibration consists of the collection of a sample of a chemical solution that is being injected to cover the desired area during the irrigation cycle. During the chemigation process, the rate of injection of a chemical compound can be calculated with the following equation:

$$g = (f \times A)/(c \times t_2 \times t_r) \qquad (8)$$

where g is the rate of injection (L/h); f is the quantity of chemical compound (kg/ha); A is the irrigation area (ha); c is the concentration of the chemical in the solution (kg/L); t_2 is the chemigation time (h); and t_r is the irrigation duration (h).

1.6   SAFETY CONSIDERATIONS IN CHEMIGATION[1,2,6,7]

Safety during the chemigation process is of paramount importance. It is recommended to use specialized equipment to protect the water source and the operators of the system and to avoid risks, health hazards, and accidents. It is necessary to install a safety valve (check valve or one-way valve) in the main line between the irrigation pump and the injection point. A manual gate valve is not enough to avoid contamination of the water source. The safety valve allows the water flow in the forward direction. If properly installed, it avoids backflow of chemicals toward the water source. It is safer to install an air relief valve (vacuum breaker) between the irrigation pump and the safety valve.

The air relief valve allows air to escape from the system. It will avoid suction of chemical solution into the water source. The pump to inject chemicals and the irrigation pump can be interconnected. The pumps are interconnected in such a way that if one is shut off, the other will shut off automatically. This is convenient when the two pumps are electrical. A safety valve must be installed in the line of chemical injection. This arrangement avoids backflow toward the tank. This backflow can cause dilution of chemical, causing spills and breakage of the system.

The spills of pesticides are extremely dangerous because these can contaminate the water source and can cause health hazards. It is recommended to locate the chemical tank away from the water source. In the case of deep well, the pesticides can wash through the soil and contaminate the well. In addition, the operator and the environment are exposed to the danger of the contamination. The safety valve generally has spring and requires pressure so that the water will flow through it. This valve allows flow only when an adequate pressure exists in the injection pump. When the injection pump is not in operation, there is no escape of liquid because of small static pressure in the tank. A gate valve at the downstream of the chemigation tank will help to avoid flow of irrigation water toward the tank when the chemigation is not in progress. This valve can be a manual gate valve, ball valve, or solenoid metering valve. The valve should be installed close to the tank. It must be open only during chemigation. Also, it must be corrosion resistant.

The automatic solenoid valve is interconnected electrically to the injection pump. This interconnection allows automatic closing of the valve in the supply line of chemical. Thus, it avoids flow of water in both directions when the chemigation pump is not in operation. The top of the chemical tank should be provided with wide openings, which will allow easy filling and cleaning. In addition, the tank must be provided with a sieve or a filter. The tank must be corrosion resistant such as stainless steel or reinforced plastic with fiber glass. The flow rate of the tank should correspond to the capacity of the pump. The tank should be equipped with an indicator to register the level of the liquid. The centrifugal pump provides high volume at low pressure. The pumps of piston and diaphragm provide volumes between moderate and high flows at high pressure. The pumps of roller and gear type provide a moderate volume at low pressure. If a pump is allowed to operate dry, then it can be damaged. It is recommended to follow the instruction manual of the manufacturer for a long life of the pump. Maintain all protectors in place. The injection pump of low volume can inject the concentrated formulation of pesticide. In that way, the problem of constantly mixing the solution in the tank is avoided. In addition, the calibration becomes easier.

One must select the hoses and synthetic or plastic tubes that can resist the operating pressure, climatic conditions, and the solvents in some chemical compounds. Do not allow the bending or kinking of hoses and tubes with another object. Wash the exterior and interior of hoses frequently so that these can last longer. These must be cleaned, washed, and stored well when are not in use. If it is possible, avoid exposure to sun. Because of climatic changes, the hoses or tubes show deteriorations on the outer surface. The areas where chemigation is in progress should display a sign that chemical compounds are being applied through irrigation system. The operator of the chemigation equipment must take all precautions: to use protective clothes, boots, protective goggles, and gloves. The waiting period to enter the field is necessary to avoid health hazards and risks. Precautions are necessary so that persons or animals do not enter the treated area during the application of pesticides and toxic substances.

1.7 EXAMPLES

Example 1
A farmer wishes to use a cloth whitener (NaOCl 1–5% active chlorine or available chlorine) to reach a concentration of 1 ppm of chlorine at the injection point. The flow rate for the system is 100 gpm. In what ratio must chlorine be injected?

$$IR = [Q \times C \times M]/S$$
$$= (100 \times 1 \times 0.006)/5 = 0.21\,gph \qquad (9)$$

where IR is the rate of chlorine injection (gph), Q is the flow rate of the system (gpm), C is the desired concentration of chlorine (ppm), S is the percentage of active ingredient (%), and M is the 0.006 for the liquid material (NaOCl) or 0.05 for the solid material $Ca[OCl]_2$.

Example 2
A farmer wants to inject Cl_2 through the drip irrigation system at a concentration of 10 ppm. What will be the rate of injection of the chlorine gas? The flow rate of the system is 1500 gpm.

$$IR = Q \times C \times 0.012 \qquad (10)$$

where IR is the rate of injection of chlorine (pound/day), Q is the flow rate of the system (gpm), and C is the desired chlorine concentration (ppm).

$$IR = 1500 \times 10 \times 0.012 = 180 \text{ pounds per day}$$

1.8 SERVICE AND MAINTENENCE OF DRIP IRRIGATION SYSTEMS

We must prevent the obstructions in the filters, laterals, and emitters. The clogging can be prevented with a good maintenance and periodic service of the system. To operate and maintain a drip irrigation system in a good working condition, the following considerations are important[2]:

1. Pay strict attention to the filtration and flushing operation.
2. Maintain an adequate operating pressure in the main, sub main, and lateral lines.
3. Flush and periodically inspect the drip irrigation system.

1.8.1 MAINTENANCE OF FILTERS AND FLUSHING OPERATION

For effective filtration efficiency, we must make sure that the system is maintained in good condition, and it is not obstructed by the clogging agents. For this purpose, pressure gauges are installed at the entrance and exit of a filter. The pressure difference between these two gauges should vary from 2 to 5 psi when the filter is clean, and the mesh is free from obstructions. The filtration system should be cleaned and flushed when the pressure difference is from 10 to 15 psi. The filters must be flushed before each irrigation operation. If water contains a high percentage of suspended solids, then the filters should be flushed more frequently. Entry of dust and foreign material should be avoided when the filters are open. Filters may not be able to remove the clay particles and algae.

1.1.1.1 FLUSHING METHOD

The frequency of flushing depends on the water quality. For flushing of irrigation lines, the following procedure can be adopted:
1. Open the ends of the distribution and lateral lines. Allow the flow of water through the lines until all the sediments are thrown out of the lines.
2. Close the ends of the distribution lines. Begin to close the lines one after another, from one block to second, and so on. There must be a sufficient pressure to flush out all the sediments.

1.8.2 CLEANING WITH PRESSURIZED AIR

The clogging can be caused by the presence of organic matter in water. It may be necessary to use pressurized air to clean the drippers. Before beginning this process, water is passed through the lines for a period of 15 min. When adequate operating pressure has been established, the air at 7 bars of pressure is allowed through the system. The compressed air will clean the lines, laterals, and drippers of the accumulated organic matter.

1.8.3   CLEANING WITH ACIDS AND CHLORINE

The clogging may also be caused by the precipitation of salts. The cleaning with acids will help to dissolve the chemical deposits. This process is not effective to remove the organic matter. Sodium hypochlorite (at the rate of 1 ppm) can be injected on the suction side of a pump for 45–90 min before shutting off the pump. The best time of injection is after flushing the sand filters, because chlorine prevents the growth of bacteria in the sand. The surface water containing iron can be treated with chlorine or commercial bleaching agent for 45 min for lowering the pH to <6.5. At pH > 6.5, certain reactions in combination with the precipitates of iron may gradually obstruct the irrigation lines.

One may use commercial grade phosphoric acid or hydrochloric acid. Before using the acid, water is allowed to pass through the system at a pressure greater than the operating pressure. Fill the fertilizer tank up to two-thirds of its capacity. Add the acid at a rate of 1 L/m³/h of flow rate. Inject the diluted acid into the system, as one will inject the fertilizer, in a normal process.

Remember: When using the chemigation tank, first pour water and then add the acid.

1.8.4   METHODS TO REPAIR TUBES OR DRIP LINES

1. The orifices of bi-wall tubing may be obstructed by salts. Polyethylene tubing of small diameter is used as a bypass method to repair these drip lines.
2. If the line is broken or if there is an excessive escape of water, the pipe or the tube is cut down and is connected with a union or a coupling.
3. If the main line is made of flexible nylon flat and is leaking, then use a small piece of plastic pipe of the same diameter to insert into the flexible nylon tubing. Both ends are sealed with the use of pipe clamps.

1.8.5   SERVICE BEFORE THE SOWING SEASON

1. Clean and flush all the distribution system and the drip lines with water.
2. Wash with water and clean the pump house system. Lubricate all valves and accessories.
3. Turn on the pump and activate the system. Check the pipes and drip lines for leakage. Repair if necessary.
4. If the system has been used previously, then cleaning and flushing should be carried out for a longer period of time. It is particularly important in sandy soils, as the sand can penetrate into the pipe during the removal of lines.

1.1.6 SERVICE AT THE END OF CROP SEASON

At the end of a crop season, following steps should be taken:
1. Flush the pipes. Clean the filters and other components of the system.
2. Lubricate all the gate valves and accessories.
3. If the pipes are permanently installed in the field and cannot be removed at the end of a crop season, keep these free of soil and weeds that can grow nearby.
4. If the system can be moved from one place to another (according to the season), the following procedure is adequate:
   a. Flush and clean the system.
   b. Remove the drip lines and collect these carefully.
   c. It is best to leave the main lines in place. If it is not possible or if there is a need for a transfer to an area, then these should be rolled. Close both ends and store in the shaded area.
   d. It is advisable to label the hoses with tags. Distance between orifices and frequency of use should be indicated on the tag.

1.9 TROUBLE SHOOTING

1.9.1. ACIDIFICATION AND CHLORATION

Cause	Remedy
Uniformity of Application Is Not Adequate	
1. Drippers are clogged with precipitates or clay particles.	Replace drippers. Inject HCl according to the instructions. Use dispersing agents such as Na and Al.
2. Drippers are clogged with microorganisms.	Use biocides, algicides, and bactericides.
3. Lines are clogged.	Flush the lines by opening the ends.
4. Filters are clogged.	Clean the filters. Open the corresponding gate valves.
Chemical Tank Is Overflowing	
5. Gate valve between two injection points is closed. Gate valve of injection line is closed.	Open the valves.
6. Filters are totally obstructed.	Flush filters.
Signs of Chlorosis	

7. Adequate dosage of fertilizers is not used.	Use recommended dosages.
8. Lack of uniformity of application.	Improve uniformity. Flush and change drippers if necessary.
9. Formulations of nitrogen that precipitate.	Use acid to clean the lines. Do chemical analysis to check the calibration.
Components Are Leaking	
10. Corrosion of the components.	Use anticorrosive components.
11. Purple color in young leaves: loss of phosphorus by precipitation in line.	Use formulations of phosphorus that do not precipitate. Apply phosphorus in bands not through the irrigation system.

1.9.2　Service and Maintenance

Causes	Remedies
Pressure Difference > Recommended Value	
1. Filters are obstructed.	Flush the filters.
2. Lines are broken.	Repair or replace lines.
3. Pump is defective.	Repair or replace the pump.
4. Gate valve is blocked.	Fix or replace the gate valve.
5. Pressure regulator is defective.	Remove and replace the regulator.
Laterals (or Drip Lines) and Drippers Are Clogged	
6. Sand is being accumulated in the drippers and lines.	Open ends of laterals and leave open for more than 2 min so that water at pressure passes through.
7. Formation of algae and bacteria.	Wash with chlorine. Paint the polyvinyl chloride pipes or install the lines below soil surface.
8. Sediments are being accumulated.	Wash with acid.
9. Precipitation of chemical compounds by chemigation.	Wash with acid and conduct the chloration process.
10. Obstruction by nest of insects.	Wash with insecticide.
Pressure Is Increased	
11. Orifices in the drip lines or drippers are clogged.	Flush the drip lines or laterals.

1.10 SUMMARY

The chemigation is an application of chemicals through the irrigation system. The chemigation provides uniformity in the application of chemicals, allowing the distribution of these chemicals in small quantities during the growing season when and where these are needed; reduces soil compaction and the chemical damage to the crop; reduces the quantity of chemicals used in the crop and the hazards and risks during the application; reduces the pollution of the environment; and reduces the costs of manual labor, equipment, and energy. The chemical injection unit consists of chemigation pump, pressure regulator, gate valve, pressure gauge, connecting tubes, check valve, and chemigation tank. All fertilizers for the chemigation must be soluble.

Chlorine can be introduced at low concentrations at necessary intervals or at high concentrations for a few minutes. Chlorine can be injected in the form of chlorine powder and chlorine gas. All chemigation equipments must be corrosion resistant, reliable, and precise. The chemigation procedure must follow a preestablished order. Irrigation system is operated until the soil saturates to a field capacity. Then, the chemicals are injected. Once the chemigation is finished, water is allowed to flow free in the system for a sufficient time to remove all the sediments from the laterals and drippers. "Rule of thumb is to chemigate during the middle of the irrigation cycle." Maintenance is a routine procedure. One should inspect all the components of the injection system after each application. It is recommended to replace the defective components. Calibration consists of adjustment of the injection equipment to supply a desired amount of chemical. It is recommended to use specialized equipment to protect the water source and the operators of the system and to avoid risks, health hazards, and accidents.

Chloration is a process of injection of chlorine compounds through an irrigation system to prevent obstruction. The chloration solves the problem of obstruction of the emitters or drippers caused by biological agents effectively and economically. The chloration can be continuous or in intervals, depending on the desired results. The most common chlorine sources are sodium and calcium hypochlorite and chlorine gas. NaOCl is safer than Cl_2. The chapter discusses the quality of water, principle of chloration, commercial sources of chlorine, methods of chloration, and examples to calculate the injection rates.

The orifices in the drip lines or the emitters emit water to the soil. The emitters allow the discharge of only a few liters or gallons per hour. The emitters have small orifices, and these can be easily obstructed. For a trouble-free operation, one should follow these instructions: pay strict attention to filtration and flushing operation. Maintain an adequate operating pressure in the main, sub main, and lateral lines. Flush and periodically inspect the drip irrigation system.

For effective filtration efficiency, we must maintain the system in good condition, and it should not be obstructed by the clogging agents. For this, pressure gauges are installed at the entrance and exit of a filter. The frequency of flushing depends on the water quality. Some recommendations for an adequate maintenance are cleaning with pressurized air, acids, and chlorine. This chapter includes methods to repair tubes or drip lines. Also, there is a procedure for service before the sowing season and service at the end of the crop season.

KEYWORDS

- acid treatment
- acidification
- air relief valve
- atmosphere
- automatic solenoid valve
- backflow
- bar
- biological agents
- centrifugal pump
- check valve
- chemigation
- chloration
- chlorination or chloration
- chlorosis
- clay
- clogging
- dissolve
- distribution lines
- drip irrigation
- dripper
- emitter
- fertigation
- fertilizer
- fertilizer tank
- field capacity
- filter
- filter, sand

- filtration system
- flow metering valve
- fungigation
- gate valve
- glycerophosphates
- herbigation
- injection system
- insectigation
- line, distribution
- line, main
- maintenance
- manual gate valve
- nemagation
- nitrogen
- nitrogen fertigation
- nutrients
- organic matter
- orifices
- pH
- phosphorus
- photosynthesis
- polyvinyl chloride (PVC) pipe
- polyethylene (PE)
- potassium
- precipitation
- pressure regulator valve
- pump
- pump house
- root system
- root zone
- sand
- slime
- sodium hypochlorite

- soil moisture
- solenoid valve
- solid particles in suspension
- solubility
- term
- union
- urea
- venturi system
- water quality
- water source
- weed

REFERENCES

1. BAR-RAM Irrigation Systems. *Instructions for the Operation and Maintenance*. p 3–5.
2. Goyal, M. R., Ed. *Chapters 8, 9 and 11*. In *Management of Drip/Trickle or Micro Irrigation*; Apple Academic Press Inc: Oakville, ON, Canada, 2013; p 165–194, 213–218. ISBN 9781926895123. © Copyright 2013.
3. Rural Development Center of University of Georgia. Proceedings of the National Symposium on Chemigation, August (1982); Rural Development Center of University of Georgia: Tifton, CA, 1982.
4. US Government – EPA [Environmental Protection Agency]. *Apply Pesticides Correctly – User's Guide for Commercial Application of Pesticides [Aplique los plaguicidas correctamente: Guía para usuarios comerciales de plaguicidas,* Spanish]; US Government – EPA [Environmental Protection Agency]: Washington, DC, 1967
5. Harrison, D. S. *Injection of Liquid Fertilizer Materials into Irrigation Systems*; University of Florida: Gainesville, FL, 1974 p 3–6, 10–11.
6. Rolston, D. E.; Rauschkolb, R. S.; Phene, C. J.; Miller R. J.; Uriu, K.; Carlson, R. M.; Henderson, D. W. *Applying Nutrients and Other Chemicals to Trickle Irrigated Crops*; University of California, 1979 p 3–12.
7. Smajstrla, A. G.; Harrison, D. S.; Good J. C.; Becker, W. J. *Chemigation Safety*; Agricultural Engineering Fact Sheet, Florida Cooperative Extension Service, Institute of Food and Agricultural Sciences, University of Florida, 1982.
8. Boswell, M. J. *Chapter 3*. In *Micro Irrigation Design Manual*; James Hardie Irrigation (Now Toro Irrigation): Fresno, CA, 1985; p 1–15
9. Goyal, M. R.; Rivera, L. E. *Chemigation Through Drip Irrigation System [Riego por Goteo: Quimigación,* Spanish]. Bulletin IA61 Series 4; Cooperative Agricultural Extension Service, University of Puerto Rico–Mayaguez Campus, 1984; p 1–15
10. Nakayama, F. S.; Bucks D. A. *Trickle Irrigation for Crop Production*; Elsevier Science Publisher B.V.: The Netherlands, 1986. p 142–157, 179–183.
11. Abbott, J. S. Emitter clogging: causes and prevention. *ICID Bull.* 1985, 34(2), 17.
12. Chlorine Kit; Taylor Chemicals Inc.: Baltimore, MD.

13. Goyal, M. R.; Rivera, L. E. *Service and Maintenance of Drip Irrigation System* [*Riego por Goteo: Servicio y Mantenimiento*, Spanish]. Bulletin IA64 Series 5; Cooperative Agricultural Extension Service, University of Puerto Rico–Mayaguez Campus, 1985 p 1–18

14. Rivera, L. E.; Goyal, M. R. *Filtration system for drip irrigation* [*Riego por Goteo: Sistemas de Filtración*, Spanish]. Bulletin IA66 Series 7; Cooperative Agricultural Extension Service, University of Puerto Rico – Mayaguez Campus, 1986 p 1–42

15. Ford, H. W. Water Quality Test for Low Volume Irrigation; Lake Alfred AREC-Florida Research Report CS79–6 HWF100, Florida, FL, 1980.

16. Ford, H. W. A Key for Determining the Use of Sodium Hyper Chlorite to Inhibit Iron and Slime Clogging of Low Pressure Irrigation Systems in Florida; Lake Alfred AREC Research Report CS79–3, Florida, FL, 1979.

17. Ford, H. W. Estimating Chlorine Requirements; Lake Alfred AREC Research Florida Report CS80–1, Florida, FL, 1980.

18. Ford, H. W. The Use of Chlorine in Low Volume Systems Where Bacterial Slimes are a Problem; Lake Alfred AREC Research Report CS75–5, HWF100: Florida, FL, 1980.

19. Ford, H. W. Using a DPD Test Kit for Measuring Free Available Chlorine; Lake Alfred AREC Research Florida Report CS79–1, Florida, FL, 1980.

CHAPTER 2

EMITTER CLOGGING: PRINCIPLES, PRACTICES, AND MANAGEMENT

VISHAL K. CHAVAN, S. K. DESHMUKH, and M. B. NAGDEVE

CONTENTS

2.1   INTRODUCTION

Monsoon rains are unevenly distributed in space and time and not adequate to meet the moisture requirement of the crops for successful farming. India, having only 4% of the total available freshwater in the world, supports about 17% of the world's population. The agricultural sector consumes more than 80% of the available water in India for the irrigation of crops and would continue to be the major water-consuming sector because of intensive agriculture.[1-3] Indian population is estimated to be 1.4 billion by 2020 with food requirement of 280 million-tons; therefore, the agricultural sector must grow by 4% and augment by about 3–4 million-tons of food per year. Although the ultimate irrigation potential of the country has been assessed at 140 million-ha in 2050, even after achieving the same, approximately half the cultivated land would still remain rain-fed. Therefore, water would continue to be the most critical resource limiting the agricultural growth.

The water resources of the country are varied and limited, but still most of the area is irrigated using the conventional methods of irrigation with an efficiency of 35–40%. Considering the daunting task of achieving food production targets, it is imperative that efficient irrigation methods such as drip and sprinkler irrigation systems are adopted on large scale for judicious use and management of water to cope up with increasing demand for water in agriculture in order to enhance and accelerate the agricultural production in the country.

Drip irrigation, also called trickle or daily irrigation, is a point-source irrigation method that slowly and frequently provides water directly to the plant root zone[4] and is the most efficient irrigation method with an application efficiency of more than 90%. However, not until the innovation of polyethylene plastics in the 1960s did drip irrigation begin to gain momentum. Traditionally, irrigation had been relied upon for a broad coverage of water to an area that may or may not contain plants. Promoted for water conservation, drip irrigation does just the opposite. It applies small amounts of water (usually every 2 or 3 days) to the immediate root zone of plants. In drip irrigation, water is delivered to individual plants at a low pressure to specific zones in the landscape or garden. The slow application promotes a thorough penetration of water to individual plant root zones and reduces potential runoff and deep percolation. The depth of water penetration depends on the length of time the system is allowed to operate and the soil texture.

The suitability of any irrigation system mainly depends upon its design, layout, and performance. Because of its merits and positive effects, drip irrigation has become rapidly popular in India, and also the state governments are promoting drip irrigation on a large scale by providing subsidy. The advantage of using a drip irrigation system is that it can significantly reduce soil evaporation and increase water use efficiency by creating a low, wet area in the root zone. World over, studies indicate that drip irrigation results in 30–70% water saving, and yield increases by about 40–100% or even more compared with surface irrigation methods. Because of water shortages in many parts of the world today, drip irrigation is becoming quite popular.[5,6] In 2000, more than 73% of all agricultural fields in Israel and 3.8 million-ha worldwide were irrigated using drip irrigation systems.[7] By 2008, total world agriculture area was 1628 million-ha of which 277 million-ha were under irrigation and 6 million-ha were drip irrigated.[8] In India, there has been a tremendous growth in the area under drip irrigation during the past 15 years. In India, the area under drip irrigation increased from a mere 1500 ha in 1985 to 70,859 ha in 1991–1992, and at present, around 3.51×10^5 ha area is under drip irrigation with the efforts of the Governments of India and the states. The *National Committee on Plasticulture Applications in Horticulture*, Ministry of Agriculture, Government of India (GoI), has estimated that a total of 27 million-ha area in the country has the potential of drip irrigation application; thus, there is a vast scope for increasing the area under drip irrigation.[9]

Because of the limited water resources and environmental consequences of common irrigation systems, drip irrigation technology is getting more attention and playing an important role in agricultural production, particularly with high-value cash crops such as greenhouse plants, ornamentals, and fruits. Therefore, use of drip irrigation systems is rapidly increasing around the world. Despite its advantages, in drip irrigation system, emitter clogging is one of the major problems that can cause large economic losses to the farmers. Emitter clogging is directly related to the quality of the irrigation water, which includes factors such as suspended solid particles, chemical composition, and microbes, and also insects and root activities within and around the tubing can cause problems. The major operational difficulties in drip irrigation method arise from the clogging of dripper, which reduces the efficiency and the crop yield.[10]

Emitter clogging continues to be a major problem in drip irrigation systems. For high-valued annual crops and perennial crops, where the longevity of the system is especially important, emitter clogging can cause large economic losses. Even though information is available on the factors causing clogging, control measures are not always successful. These problems can be minimized by appropriate design, installation, and operational practices. Reclamation procedures to correct clogging increase maintenance costs and, unfortunately, may not be permanent. Clogging problems often discourage the operators and consequently cause abandonment of the system and return to a less efficient irrigation application method.

Emitter clogging is directly related to the quality of the irrigation water, which includes factors such as suspended particle load, chemical composition, and microbial type and population. Insect and root activities within and around the tubing can also cause similar problems. Consequently, these factors dictate the type of water treatment or cultural practices necessary for clogging prevention. Clogging problems are often site specific, and solutions are not always available or economically feasible.[11] No single foolproof quantitative method is available for estimating the clogging potential. However, by analyzing the water for some specific constituents, possible problems can be anticipated and control measures be formulated.

Most tests can be conducted in the laboratory. However, some analyses must be done at the sampling sites because rapid chemical and biological changes can occur after the source of water is introduced into the drip irrigation system. Water quality can also change throughout the year so that samples should be taken at various times over the irrigation period. These are further rated in terms of an arbitrary clogging hazard ranging from minor to severe. Clogging problems are diminished with lower concentrations of solids, salts, and bacteria in the water. Additionally, clogging is aggravated by water temperature changes.

The causes of clogging differed based on emitter dimension[12,13] and positions in lateral. The tube emitter system with laminar flow suffered more severe clogging than the labyrinth system with turbulent flow, because laminar flow is predisposed to clogging.[14] Emitter clogging was recognized as inconvenient and one of the most important concerns for drip irrigation systems, resulting in lowered system performance and water stress to the nonirrigated plants.[15] Partial and total plugging of emitters is closely related to the quality of the irrigation water, which occurs as a result of multiple factors, including physical, biological, and chemical agents.[16,17] Favorable environmental conditions in drip irrigation systems can cause rapid growth of several species of algae and bacteria resulting in slime and filament buildup, which often becomes large enough to cause biological clogging.[18] On the other hand, some of the bacterial species may cause emitter clogging due to the precipitation of iron, manganese, and sulfur minerals dissolved in the irrigation water.[17,19] Filtration, chemical treatment of water, and flushing of laterals are means generally applied to control emitter clogging.[20,21] Physical clogging can be eliminated with the use of fine filters and screens. Emitter clogging is directly related to irrigation water quality, which is a function of the amount of suspended solids, chemical constituents of water, and microorganism activities in water. Therefore, the above-mentioned factors have a strong influence on the precautions that will be taken for preventing the plugging of the emitters. During irrigation, some clogging due to microorganism activities takes place in cases when wastewater is used.[22,23]

In micro irrigation systems that are characterized by a number of emitters with narrow nozzles, irrigation uniformity can be spoilt by the clogging of the nozzles with particles of chemical character.[24,25] Chemical problems are due to dissolved solids interacting with each other to form precipitates, such as the precipitation of calcium carbonate in waters rich in calcium and bicarbonates.[26] In locations where the amount of the ingredients such as dissolved calcium, bicarbonate, iron, manganese, and magnesium are excessive in irrigation water, the emitters are clogged by the precipitation of these solutes.[27] Chemical precipitation can be controlled with acid injection. However, biological clogging is quite difficult to control. Chlorination is the most common practice used in the prevention and treatment of emitter clogging caused by algae and bacteria.[28,29] Calcium hypochlorite, sodium hypochlorite, and particularly chlorine are the most common and inexpensive treatments for bacterial slimes and for the inhibition of bacterial growth in drip irrigation systems.[4,30,31]

However, continuous chlorination would increase total dissolved solids in the irrigation water and would contribute to increased soil salinity.[32]

In this chapter, the past research studies are reviewed, presented, and discussed to identify the causes of clogging, water quality impacts, and preventive measures, and that related to adsorption mechanism in physical, chemical, and biological clogging of drip irrigation.

2.2 MECHANISM OF EMITTER CLOGGING: A REVIEW

2.2.1 NATURE AND SCOPE OF PRESENT STATUS

The various components of drip irrigation are made up of plastic and polymer materials because of their flexibility and other advantages over metals. There are more than 200 various components and materials used in drip irrigation installed at farm level. But the major components coming in contact with water includes poly tube (material: linear low-density polyethylene (LLDPE)), dripper (material: polypropylene (PP)), pipes (material: high-density polyethylene (HDPE) and polyvinyl chloride (PVC)), and silicone diaphragm in emitters (material: silicone). It is found that the initiation of clogging is at molecular level. Considering this, it is very important to study the clogging mechanism and initial adsorption mechanism in drip irrigation system. The phenomenon of adsorption is the collection or accumulation of one substance on the surface of another substance. In adsorption, mainly the surface of solids is involved, and accumulated substances remain on the surface. Adsorption therefore is said to be a surface phenomenon as it occurs because of attractive forces exerted by atoms or molecules present at the surface of the adsorbent. These attractive forces may be of two types: (i) physical forces (cohesive forces or van der Waals forces) and (ii) chemical forces (chemical bond forces). Thus, an attempt on the study of adsorption parameters of these materials of silicone diaphragm, PVC, PP, HDPE, and LLDPE used in drip irrigation would give insight into which material is more susceptible for adsorption and also possible solutions to reduce clogging mechanism in the initial stage itself.

2.2.2 CAUSES OF CLOGGING

The major disadvantage and the only real concern with drip irrigation systems is the potential for emitter clogging. The causes of clogging fall into three main categories: (1) physical (suspended solids), (2) chemical (precipitation), and (3) biological (bacterial and algal growth). Emitter clogging is usually the result of two or more of these elements working together.[33] Emitter clogging is a major concern in these systems because of the high suspended solids and nutrient values associated with treated wastewater effluent. Early research on drip irrigation systems investigated two different but related issues regarding emitter clogging. One focused on the clogging mechanism of emitter and the other focused on the adsorption mechanism in emitters.

In order to avoid emitter clogging, one could use larger emitter orifices, and to maintain the application rate at optimum, the system had to be operated by impulse methods.[34] To prevent clogging problem due to bacteria or algae, addition of chlorine on daily basis with a concentration of 10–20 mg L^{-1} is recommended. Precipitates from bacterial activity or algae growth will also be removed if chlorine is injected with high concentration (500 mg·L^{-1}) inside the system for 24 h before washing.[35] Clogging problems are often site specific, and the solutions are not always available or economically feasible.

2.2.2.1 PHYSICAL CLOGGING

The biological plugging of emitter is a problem in drip irrigation system installed in various citrus groves in central and south Florida. They found that the clogging problems in drip irrigation system were due to isolation of algae in the irrigation lines and emitters.[36]

The micro tube emitters clogged more rapidly than any other type of emitters, and the chemical treatments were effective on emitters.[37] They further conducted a study to test the effect of water temperature variation upon the discharge rates of micro tube, spiral passage, orifice, and vortex-type emitters.

The dominant causes of emitter clogging and observed flow reduction were physical factors, viz., sand grain, plastic particles, sediment, body parts of insects and animals, deformed septa, which contributed to 55% of the total clogging that occurred during their study, followed by the combined development of biological and chemical factors, such as microbial slime, plant roots and algae mats, carbonate precipitate, iron–magnesium precipitates, which contributed to 16% of the total clogging.[38] They tabulated the dominant causes of clogging with their relative % occurrence in trickle irrigation emitter as shown in Table 2.1.

TABLE 2.1 Dominant Causes of Clogging and Their Relative Percentage of Occurrence in Drip Irrigation Emitters[20]

Causes of Logging[20]	Percentage of Occurrence	
	Individual	Total
Physical factors		
Sand grains	17	
Plastic particles	26	
Sediment	02	
Body parts of insects and animals	03	
Deformed septa	07	55

TABLE 2.1 *(Continued)*

Causes of Logging[20]	Percentage of Occurrence	
	Individual	Total
Biological factors		
Microbial slime	11	
Plants root and algae mats	03	14
Chemical factors		
Carbonate precipitate	02	
Iron–magnesium precipitate	00	02
Combined factors		
Physical/biological	08	
Physical/chemical	02	
Chemical/biological	06	
Physical/biological/chemical	02	18
Nondetectable (probably physical)	–	11

A field study evaluated the effects of emitter design and influent quality on emitter clogging using eight different emitters in combination with six water treatments of Colorado River water, USA.[39] The treatments were varied by the degree of filtration for the removal of suspended solids and the amount of chemical additives (sulfuric acid and calcium hypochlorite) for controlling pH and preventing biological growth. They found that the emitter performance was reduced during average flow rates for each experimental subplot. They recommended a combination of sand and screen filtration to remove suspended solids, acid treatment to reduce chemical precipitation, and flushing of laterals to eliminate sediments.

Emitter clogging of particulate removal by granular filtration and screen filtration of wastewater stabilization pond effluent has been discussed by Adin.[40] He showed that deep-bed granular media controlled suspended particles in which larger particles (10 μm) were removed at a higher rate than smaller particles, and very little removal occurred in the 1–2 μm size range; further 10–60 μm size particles were removed by 40–50% "in-depth filtration" and by more than 80% when bio-mats developed and surface filtration occurred. He found that the particulate removal efficiency increased as the filter media size approached 1.20 mm and the filter bed depth increased to 0.45 m and decreased (especially for small particles) with increased filtration velocity from 2.2 $L \cdot m^{-2} \cdot s^{-1}$. He also observed that the release of large particles from the screen into the effluent during the formation of "filter cake" actually increased the chance of clogging in the irrigation drip lines and emitters.

He suggested the manufacturers of wastewater drip irrigation systems on numerous modifications to emitter design and other system components to incorporate the suggestions of early researchers to reduce clogging and maintain distribution uniformity.

An experiment was conducted at Farm Research Institute in China on subsurface drip irrigation (SDI) system related to emitter clogging for 8 years.[41] The clogging rates of labyrinth emitter, mini-pipe, and orifice reached 16.67, 25, and 63.89%, respectively. Clogging was found to be mainly caused by attached granules, and some suggestions they put forward to solve this problem were enhancing filtration, flushing the filter timely, and improving the route of water in the emitter. Causes of clogging are listed in Table 2.1.

The efficiency of using nitric acid and sodium hypochlorite was evaluated in drippers with clogging caused by the use of water with high algae content in a rose field of Holambra, Sao Paulo State, Brazil.[42] Researchers conducted the research study in six, 4216 m^2 greenhouses, each with two sectors comprising 10 spaces or lines, totaling 12 sectors of a dripper irrigation system. Their results on irrigation water quality in the greenhouses indicated that the pH and iron presented moderate risk for clogging. Chemical and physical analyses and the bacteriological count in water were carried out in three water sources that supplied the irrigation water to check the factors causing clogging. Evaluations were carried out on water distribution uniformity in all sectors before and after the chemical treatment in order to evaluate efficiency. The treatment improved water distribution uniformity and led to a reduction in the coefficient of variation for dripper flow in all sectors. There was a good correlation between the coefficient of variation (Cv) and the water distribution uniformity index. According to them, this was an excellent method to be used in drippers with clogging due to biological problems.

A laboratory experiment was conducted to evaluate emitter clogging in drip irrigation by solid–liquid two-phase turbulent flow model simulations describing the flow within drip emitters.[43] The moving trace and depositing feature of suspending solids in emitter channels by computational fluid dynamics (CFD) were based on the turbulent model established, which provided some visual and direct evidences for predicting the clogging performance of drip emitters. Three types of emitters were used with novel channel form, including eddy drip-arrows, pre-depositing drippers, and round-flow drip-tapes. The simulation results showed that the solids moved along a helical path in the eddy drip-arrow, but no obvious deposition existed in its interior channel. In the pre-depositing dripper, some solids concentrated in the parts of "depositing pones". In the round-flow drip-tape, a small number of solids adhered to outer edges of every channel corner, which was a potential factor for the occurrence of emitter clogging. To verify the predictions from the CFD simulations, a series of "short-cycle" clogging tests for the three emitters were conducted in laboratory. The statistical data of discharge variation caused by emitter clogging were in good agreement with the two-phase flow CFD simulations.

2.2.2.2   CHEMICAL CLOGGING

At the Tamil Nadu Agricultural University, the average clogging was found to be 20–25% in the nozzles, 34% in the micro tube, and 40% in hole and socket type of emitters.[44] The study concluded that the extent of reclamation of partially clogged emitters treated with hydrochloric acid resulted in an increase of 0.2 $L \cdot h^{-1}$ when acid concentration increased from 0.5 to 2% by volume.

A field experiment on drip irrigation was installed in gold coast farms of Santa Maria near coastal and southern part of California.[45] Researchers divided a 10-acre field of strawberry into six plots. They used the maleic anhydride polymer injected at 2 $mg \cdot L^{-1}$ and continuous chlorine of 1 $mg \cdot L^{-1}$ in four plots, and used continuous chlorine injection only in two plots. They found that emission evaluation of polymer-treated water showed only slight decrease over the 6-month growing period, but untreated well waters indicated a decrease of 50%. They finally concluded that the system injected with 2 $mg \cdot L^{-1}$ maleic anhydride polymer supplied the actual amount of water required for the plant needs, but the untreated tubings output decreased.

The effects of chemical oxidants on effluent constituents in drip irrigation were evaluated with synthetic effluents rather than authentic effluents in order to realize the role of oxidants in these processes and to obtain meaningful and reproducible results.[46] Researchers studied the effects of Cl_2 and ClO_2 and their constituents. The demand of these effluents was 5–8 $mg \cdot L^{-1}$ for Cl_2 and 3–4 $mg \cdot L^{-1}$ for ClO_2. They found that 2 $mg \cdot L^{-1}$ of either oxidant caused a very fast bacteria inactivation, which reached four orders of magnitude after 1 min. However, with respect to algae, concentrations up to 20 $mg \cdot L^{-1}$ of either oxidant did not affect the number of algae cells, although they caused a remarkable decrease in algal viability as expressed by its chlorophyll content and replication ability. Both oxidants demonstrated a notable aggregation effect on the effluents. The conclusions of the results described above were examined in a pilot system. They suggested that continued chlorination by 5–10 $mg \cdot L^{-1}$ Cl_2 applied directly to the drippers was not very effective. The reason for this was the presence of clogging agents, "immune" to low Cl_2 concentrations, produced as early as in the reservoir, and carried down to the drippers by the effluent stream. Finally, they suggested that batch treatment combined with settling was much more efficient as it reduced clogging significantly, because in this case, Cl_2 reacted not only as a disinfectant, but also as a coagulant due to the oxidation of humic constituents.

Enciso and Porter[47] worked on a maintenance program that included cleaning the filters, flushing the lines, adding chlorine, and injecting acids. They reported that if these preventive measures were done, the need for major repairs, such as replacing damaged parts, often can be avoided and the life of the system be extended.

Precipitation of salts such as calcium carbonate, magnesium carbonate, or ferric oxide can cause either partial or complete blockage of the drip system.[48] Acid treatment was applied to prevent the precipitation of such salts in drip irrigation system, and it was also found effective in cleaning and unclogging the irrigation system, which was already blocked with precipitates of salts.

The past research suggested that the irrigation pumping plant and the chemical injection pump should be interlocked because if the irrigation pumping plant were to stop, the chemical injection pump would also stop. This would prevent chemicals from the supply tank from entering the irrigation lines, when the irrigation pump stops.[20]

The drip system should be flushed as regularly as determined by water quality, monitoring, and recording system. Flushing process should start from the pump onward to make sure that the filters are clean and pressure is set correctly, and systematically, the mainline, sub-mains, laterals, and flushing manifold should be cleaned.[49]

Five types of emitters with different nominal discharges were evaluated with or without self-flushing system and with or without pressure-compensating system under three management schemes: untreated well water, acidic treated water, and magnetic treated water in order to reduce chemical clogging.[50]

In order to save money, concentrated and inexpensive technical acids should be used such as concentrated commercial grade hydrochloric, nitric, or sulfuric acid.[51] Phosphoric acid, applied as fertilizer through the drip system, can also act as a preventive measure against the formation of precipitates under certain conditions.

A comparison for clogging in emitters was made during the application of sewage effluent and groundwater for investigating temporal variations of the emitter discharge rate and the distribution of clogged emitters in the system and for quantifying the impact of emitter clogging on system performance.[52] In the experiment, six types of emitters (with or without a pressure-compensation device) and two types of water sources (stored secondary sewage effluent and groundwater) were assessed by measuring the emitter discharge rate of the system at approximately 10-day intervals. The total duration of irrigation was 83 days, and there was a daily application for 12 h. The water source had a very significant influence on the level of drip emitter clogging. Of all the emitters tested over the entire period of the experiments, the emitters applying sewage effluent were clogged much more severely, producing a lower average mean discharge rate 26% than those applying groundwater. They found that different types of emitters had different susceptibilities to clogging, but the clogging sensitivity was inversely proportional to the pathway area in the emitter and the emitter's manufacturing coefficient of variability (Mfg Cv). For groundwater application, the clogged emitters tended to be generally located at the terminals of the laterals in case of emitters without a pressure-compensation device, whereas randomly distributed clogged emitters were found for pressure-compensating emitters. A more random distribution of clogged emitters was found for the sewage application. Clogging of emitters could seriously degrade system performance. They found that the values of the uniformity coefficient (CU) and the statistical uniformity coefficient (Us) decreased linearly with the mean clogging degree of emitters.

Finally, they suggested that to maintain a high system performance, more frequent chemical treatments should be applied to drip irrigation system using sewage effluent than to systems using groundwater.

2.2.2.3   BIOLOGICAL CLOGGING

2.2.2.3.1   IRON BACTERIA

Emitter clogging problem occurring in certain areas of Florida was caused by iron bacteria.[53] Iron bacteria frequently thrive in iron-bearing water. It is unclear whether the iron bacteria exist in groundwater before well construction and multiply as the amount of iron increases because of pumping, or whether the bacteria are introduced into the aquifer from the subsoil during well construction. Well drillers should use great care to avoid the introduction of iron bacteria into the aquifer during the well drilling process. All drilling fluids should be mixed with chlorinated water at 10 mg·L^{-1} free chlorine residual.

2.2.2.3.2   SULFUR BACTERIA

Florida micro irrigation systems ceased to function properly because of filter and emitter clogging caused by sulfur bacteria.[54] *Thiothrix nivea* is usually found in high concentrations in warm mineral springs and contributes to this problem. This bacterium oxidizes hydrogen sulfide to sulfur and can clog small openings within a short period. *Beggiatoa,* another sulfur bacterium, is also often found in micro irrigation systems. Continuous chlorine or copper treatment can control sulfur bacterial problems.

2.2.2.3.3   MICROORGANISMS

Clogging of micro sprinklers in a Florida citrus grove was noted to be inversely related to the orifice dimension of the emitters.[55] Clogging occurred when irrigation water from a pond was used where the water was chemically conditioned and filtered through a sand media filter. Emitter clogging was caused by algae (46%), ants and spiders (34%), snails (16%), and solid particles such as sand and bits of PVC (4%). About 20% of the 0.76 mm orifice emitters required cleaning or replacement during each quarter compared with about 14% for the 1.02 mm, 7% for 1.27 mm, and 5% for 1.52 mm orifice emitters.

   Pests can damage or clog emitters or system components and increase maintenance costs. Some pests can also cause leakage problems and others can cause clogging. Coyotes and burrowing animals were observed to damage micro irrigation tubing. Similarly, rats and mice chewed holes in micro irrigation tubing.[56] *Tortric-*

idae (Lepidoptera) larvae and pupae of *Chrysoperla externa* (Hagan) caused emitter clogging.[57] Emitter clogging by spiders and ants is a problem in many surface micro irrigation and micro sprinkler systems. Additional problems include damage to spaghetti tubing and destruction of diaphragms in pressure-compensating emitters.

2.2.2.3.4. LARVAE

The larvae of *Selenisa sueroides* (Guenee) damaged the spaghetti tubing of micro irrigation systems in south Florida citrus groves.[58] Larvae of *S. sueroides* used the spaghetti tubing on micro sprinkler assemblies as an alternate host to native plant species with hollow stems. The *S. sueroides* caterpillars chewed holes in the spaghetti tubing in order to enter and pupate. The deterioration appeared to be selective as *S. sueroides* damaged stake assemblies constructed with black tubing to a larger extent than that with colored spaghetti tubing.[59] In addition, tubing composition could also affect the severity of damage, and there was a higher incidence of damage on polyethylene tubing compared with vinyl spaghetti tubing, and even less damage on assemblies made with colored tubing.

Problems caused by the *S. sueroides* caterpillars can be controlled by methods other than pesticide treatment of the emitters. Elimination of known host plants in early autumn through herbicide application and mowing is probably more cost effective than pesticide treatment. When infestations of *S. sueroides* caterpillars persist after mid-September, spray applications of liquid Teflon® to the emitter assemblies provide some protection from *S. sueroides* larvae.

2.2.2.3.5 ANTS

Ants frequently cause clogging of emitters leading to nonuniform water application. In addition to causing clogging, ants can also physically damage certain types of emitters causing increased flow rates. For example, red fire ants (*Solenopsis invicta*, Buren) chewed and damaged the silicone diaphragms that control the flow of micropulsators. Ant activities decreased the diaphragm mass (partial to complete removal) and increased the orifice diameters. Extensive damage occurred within a 16-month period, and more than 9000 m of micro irrigation tubing with internal compensating emitters had to be replaced.[60]

2.2.2.3.6 PLANT ROOTS

Problems with plant roots affecting water flow in subsurface micro irrigation systems were observed. This was a case of "the root biting the hand that feeds it." Roots were found penetrating into cracks and grow within the tubing or the emitter and restrict water flow. A control of root activity was done by chemical injection or impregnation of the tubing, emitter, or filter with trifluralin.[61] Massive root growths were observed to physically pinch flexible tubing, thus restricting water flow.

2.2.2.3.7 BIOFILM STRUCTURE

The bacteria in a biofilm are found to aggregate at different horizontal and vertical sites with the highest concentration of cells occurring at the base or at the top, resulting in a mushroom-like shape (Fig. 2.1). The biofilms are highly hydrated with cells making up about one-third of the volume; the rest is mostly water. The horizontal and vertical voids provide for the flow of water through the network of cells. This simple type of circulatory system that carries nutrients to and waste product away from the cells is yet another example of how biofilms may function more like eukaryotic multicellular organisms. A wide variety of microenvironments exist within the cell matrix. This encourages phenotypic variations among genetically identical cells and provides for a greater diversity of species within the biofilm community.[62]

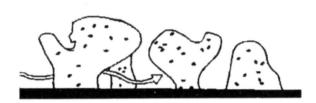

FIGURE 2.1 Biofilm structure.

2.2.2.3.8 ROLE OF BIOFILMS IN EMITTER CLOGGING

A field study of water quality and preventative maintenance procedures were conducted to reduce clogging using a variety of treatment methods and emitter systems.[63] The examination of clogged emitters concluded that three factors contributed to emitter clogging: (1) organic and inorganic suspended solids; (2) chemical precipitation of calcium, magnesium, heavy metals, and fertilizer constituents; and (3) bacterial filaments, slimes, and depositions. Authors concluded, however, that these contributors to clogging were closely interrelated. For example, by reducing biological slime, suspended solids were observed less likely to stick to the slime and agglomerate inside the tubing and emitters. Microorganisms were also found responsible for the chemical precipitation of iron, sulfur, and magnesium. For the successful operation of drip irrigation, it is recommended to use water filtration, chemical treatment, and pipeline flushing.

A research study was conducted in the field and laboratory using oxidation pond effluent aimed at defining the clogging factors and mechanisms of blockage within the emitter for developing technical measures to overcome the problem.[64] Two types of influents were tested using three different emitter types. The clogged emitters were carefully removed from the lines and were carefully opened to examine the contents. Researchers reported that slimy gelatinous deposits of amorphous shape (biofilms) served as triggers for serious emitter blockage. Particles of definable shape (inorganic solids) were found in the matrix of the gelatinous substance and formed sediment in the emitters. It was concluded that suspended solids in the influent primarily caused emitter clogging, but the clogging process was initiated by bacterial biofilms.

Effects of Treflan injection on winter wheat growth and root clogging of subsurface drippers were investigated.[65] To attempt to solve the problem of root clogging of drippers, a series of field experiments were performed in the growing seasons of 2006–2008, to investigate the effects of Treflan injection on dripper clogging by roots, and on root distribution, yield, and the quality of winter wheat (*Triticum aestivum* L.) under SDI system. For each growing season, two Treflan injection dates (March 6 and April 15 for the 2006–2007 growing season, and March 6 and April 15 for the 2007–2008 growing season) and three injection concentrations (0, 3, and 7 mg·L^{-1}) were arranged in a randomized block experimental design. The experimental results showed that Treflan injection effectively reduced root density in areas adjacent to drippers, thereby significantly decreasing the potential of root clogging. In 2007, 4 out of 35 drippers were found with root intrusion problems in the control (without Treflan injection), whereas no root clogging existed in any dripper in Treflan application treatments. In 2008, six drippers from the control but only one dripper from those treated with Treflan application showed root clogging. In addition, within the range of concentrations used by the current experiment, Treflan concentrations had no significant effects on winter wheat root distribution, yield, and quality. Injection date, however, influenced the vertical root distribution significantly. Injection of Treflan late in the growing season influenced the root distribution only in the areas close to the drippers, and the influenced areas increased if Treflan was injected early in the growing season.

2.2.2.3.9   BIOFILM DEVELOPMENT

Biofilm formation is thought to begin when environmental signals trigger the transition from a planktonic lifestyle to a sessile lifestyle. It appears that flagella22-mediated motility are important in establishing the initial cell surface contact that leads to the colonization of solid surfaces.[66] Once a cell makes contact with the surface, adhesion rather than motility is the key to successful colonization. The production of specific outer membrane proteins that provide for stable attachments is necessary for biofilm formation.[67] Once the surface is colonized with a monolayer of cells, the bacteria move across the surface organizing themselves into micro colonies. This coordinated movement called "twitching motility" was accomplished through the contraction of type IV pili. Cells within the microcolony secrete the exopolysaccharide (EPS) matrix that functions like glue holding the bacterial cells together.[68]

An experimental study was conducted to evaluate the performance of various irrigation emitters widely used in Israel, using wastewater from a storage reservoir.[69]

Fine particulate matter agglomerated by microbial by-products and inclined developed biomass were the principal clogging agents. They reported that the clogging process started with emitters located at far end of the lateral, and partial emitter clogging was more common than complete clogging.

Both field survey and experimental studies were carried out to identify the causes of clogging[70] by collecting the samples of clogging material and water for microscopic and chemical analysis in order to. Researchers found that the colonial protozoa occupied the irrigation equipment only at one location where the water flow velocity was slower than 2 m·s^{-1}, and the colonies of protozoa and sulfur bacteria were found attached to the walls of the irrigation equipment and growing in a direction perpendicular to the surface to which they were attached. Colonial protozoa developed in a wide range of water qualities: temperature (17–32°C), pH (6.8–9.5), dissolved oxygen (0.5–10.4 mg·L^{-1}), BOD (0.33–55 mg·L^{-1}). *Epystilys balanarum*, from the order *Peritricha*, was the species of colonial protozoa found. It was further concluded that the size of colonies varied from single or a few cells up to several hundred cells in a colony.

The performance was evaluated for two different types of emitters under various filtration methods as well as the effect of chemical treatments designed to control clogging using reservoir water containing secondary sewage effluents and storm water runoff.[71] The influent was filtered using a media filter of a uniform bed of gravel with a mean size of 1 mm, a 140-mesh disk filter, or a 155–200 mesh screen filter. Emitter performance was determined by automatic measurements of pressure and flow rate in each drip lateral along with manual measurements of the discharge rate from individual emitters. It was observed that the emitter clogging was associated mainly with mucous products of microbial communities including colonial protozoa, bryozoa, and bacteria.

Treated effluent was evaluated for drip-irrigated eggplant at As-Samra experimental site.[72] The study included the determination of soil characteristics prior to irrigation and the accumulation of salts and heavy metals in the soil as well as concentration of the nutrients and heavy metal accumulation in the plant tissues. Clogging of the irrigation system was evaluated and treated, and the yield was determined. The results of the study showed that the effluent had low heavy metal content and moderate restriction for surface trickle irrigation. The soil analysis indicated that after eggplant harvest, there was a slight increase in heavy metals and salt accumulation at the periphery of the wet zone. Finally, it was suggested that clogging can be successfully controlled with acid and chlorine.

A laboratory experiment in the Irrigation Department of Rural Engineering at ESALQ, USA, was performed to evaluate the use of different chlorine doses to recover the original flow rate for total and partial clogging of emitters.[73] Scientists evaluated five different types of Netafim drippers: Streamline 100, Ram 17L, Drip line 2000, Tiran 17, and Typhoon 20. Four chlorine levels of 150, 300, 450, and 600 mg·L^{-1} were tested maintaining the irrigation water pH between 5.0 and 6.0 in the recovering process of blocked drippers. A regular water source for irrigation equipment was used presenting a bacterial population of about 50,000 bacterium·L^{-1}. Sodium hypochlorite (12%) was used as a chlorine supply. The chlorination process took place for 60 min, and after this period, the solution remained inside the hose for 12 h without flow. Then, the flow of each emitter was measured. Average flow, coefficient of variation of average flow, and the percentage of the number of emitters at different classes of flow reduction were analyzed. It was concluded that for all kinds of drippers, except the Streamline, chlorine improved the flow rate.

The performance was assessed for five different drip line types receiving filtered but untreated wastewater from a beef feedlot runoff lagoon.[74] Study included five drip line types, each with a different emitter flow rate. The emitter flow rates tested were 0.15, 0.24, 0.40, 0.60, and 0.92 gal·h^{-1} per emitter. Each drip line was replicated three times in plots of 20×450 ft. Acid and chlorine were injected in drip lines and flushed as needed to prevent algae and bacteria from growing and accumulating in the drip lines. After 3 years of study, it was concluded that the three largest emitter sizes (0.4, 0.6, and 0.92 gal·h^{-1} per emitter) showed little sign of clogging and had less than (5%) reduction in flow.

Biofouling were identified as a major contributor to emitter clogging in drip irrigation systems that distributed reclaimed wastewater.[75] Two types of drip emitters were evaluated for use with reclaimed wastewater in the study. Microbial biofilm accumulations, including proteins, polysaccharides, and phospholipid fatty acids, were tested to determine the biofilm development and diversity in the emitter flow path. The microbial biofilm structure was analyzed using scanning electron microscopy. The results showed that rapid growth of the biofilm and accumulated sediments led to eventual reduction of emitter discharge, and the biofilm played an inducing role in the clogging process and biomass growth in the emitter flow path fluctuated as biomass was scoured off the surface areas. This study provided some suggestions for the control of clogging and a framework for future investigations into the role of biofilms in the clogging of drip emitters that distribute reclaimed wastewater.

2.2.3   IMPACTS OF WATER QUALITY ON CLOGGING

Emitter clogging and crop response were evaluated in a replicated field study of Netafim labyrinth emitters distributing primary and trickling filter wastewater plant effluent compared with a tap water control.[76] Emitter clogging was determined by measuring the flow rate at the end of one irrigation season. Data from this study shows that more emitters clogged at the beginning of laterals than at the ends. It may be because of the differences in pressure associated velocity heads throughout the system. It was concluded that distributing wastewater could clog emitters, which can lead to poor water distribution and severe crop damage. It was recommended that (1) the effluent be screened with automatic flushing filters; (2) chemical reagents, such as hypochlorite, be added to dissolve obstructions in the six emitters; and (3) uniform emitter clogging should be maintained along the laterals by maintaining consistent pressure heads throughout the laterals.

An experiment on trickle irrigated citrus orchard was conducted with Colorado River water in southwestern Arizona to develop water treatment methods for preventing emitter clogging and maintaining long-term operation of the system under actual field conditions.[77] The study included eight emitter systems in combination with six water treatments during a comprehensive 4-year study. Authors found that emitters with flexible membranes either failed after a few months of use with chemically conditioned water or showed serious deterioration and decomposition after 4 years. Finally, it was observed that the dominant causes of emitter clogging and flow reduction were physical particles; next and minor in comparison was the combined development of biological and chemical deposits.

An experiment was carried out at the IFAS Lake Alfred Citrus Research and Education Center for the irrigation and fertigation study of citrus.[78] In the study, authors measured monthly clogging percentages for a 5-year study of trickle irrigation of citrus. They found that clogging percentages were much greater for drip emitters as compared with spray-jets. Further, they observed that most drip emitters were clogged by iron deposits, whereas most spray-jets were clogged by insects, and drip emitters clogged much more frequently during low-use winter months as compared with high-use summer months. Finally, they concluded that clogging percentages increased annually from 2.5% in 1979 to 21.3% in 1983. Clogging percentages of spray-jets were low (<2.5%) and were unaffected by time or season throughout their study.

It has been observed that lime precipitate clogged the buried drip irrigation systems, which was difficult to detect, and could cause problems where water quality was poor.[79] On the basis of the field trials, it was suggested that injection of phosphonate (organophosphorus compounds) was as effective as acid against clogging and might cost less. The experimental procedure consisted of preparing a tank with chemical constituents appropriate to the treatment and pumping at constant pressure into four buried drip irrigation lines corresponding to the treatment. As the treatments were being irrigated, the flow rate to each of the four drip lines was monitored. The procedure was continued for each of the 16 treatments. All the treatments were irrigated for 2–3 days, and each was irrigated 10 times during the experiment. Finally, it was concluded that the treatments with low levels of calcium (1 meq·L^{-1}) showed no significant clogging problem. This study indicated that acid injection reduced clogging.

An experiment was carried out at Ft. Pierce Agricultural Research and Education Center to evaluate clogging rates for 10 models of micro irrigation emitters for a period of 3–5 years.[80] Five spray emitter models and five spinner models were used in a randomized complete block design with five replications. It was found that clogging was caused by ants, spiders, or bacteria and algae. The average clogging rate per inspection period ranged from 2 to 38% averaging 19%. It was observed that the emitter, which had a relatively large orifice and a mechanism to plug the orifice when not in use, had the lowest clogging rate.

A field study was conducted to evaluate emitter clogging under different methods of filtration for the removal of suspended solids and chlorination to prevent biofilm formation using secondary wastewater treatment plant effluent.[81] It was concluded that sand filtration and intermittent chlorination (2 mg·L^{-1} for 1 h every 22 h of operation) prevented clogging in pressure-compensating emitters. Finally, it was suggested that an intermittent chlorination of 2 mg·L^{-1} with screen filtration and a continuous chlorination of 1 mg·L^{-1} with sand filtration prevented clogging in turbulent flow labyrinth emitters.

All irrigation systems require proper maintenance, and the SDI systems were no exception.[82] The major cause of failures in the SDI and other micro irrigation systems worldwide was clogging.

Many producers use drippers for trickle irrigation systems for flower production in the field and in protected environments.[83] A frequent problem in this type of irrigation system is the clogging of drippers, which is directly related to water quality and filtering system efficiency.

2.2.4 PREVENTIVE MEASURES FOR CLOGGING IN DRIP IRRIGATION SYSTEM

An experiment in southwest Portugal was carried out to investigate the causes of emitter clogging in waste stabilization pond effluents used for drip irrigation of crops.[84] The emitter that operated most successfully utilized a long water path labyrinth to reduce flow to required level. Clogging was shown to result from the deposition and entrapment of sand particles within the emitter. It was concluded that pond micro algae alone did not constitute a major hazard to the operation of drip irrigation equipment and that waste stabilization pond effluents might be used for drip irrigation.

The effluents of different qualities for drip irrigation were examined at the Ohio State University Extension, Ohio Agricultural Research and Development Center Site.[85] The experimental emitters were designed for use with treated wastewater and contained antimicrobial agents to prevent emitter clogging. It was observed that many clogged emitters recovered to near the original flow rates after the end of the experiment.

The microbial organisms were evaluated to prevent clogging in drip irrigation system caused by biological factors.[6] In the study, three antagonistic bacterial strains were used in the *Bacillus* spp. (ERZ, OSU-142) and *Burkholderia* spp. (OSU-7) for the treatment of biological clogging of emitters. It was concluded that antagonistic bacterial strains have the potential to be used as anti-clogging agents for treatments of emitters in drip irrigation system. Finally, it was suggested that the use of antagonistic bacterial strains in drip irrigation may reduce or completely eliminate the need for repetitive chemical applications to treat emitter clogging, and these strains have the potential to be used not only for cleaning of biologically clogged emitters, but also for biological control of pathogenic microorganisms that cause diseases in plants watered with drip irrigation systems.

A study in Japan under Tohaku Irrigation Project was undertaken to reduce emitter clogging induced by biological agents such as algae and protozoa (AP) to enhance the performance of drip irrigation using chlorination.[86] The main objective of the study was to quantify the impact of AP-induced changes on discharge rate and uniformity from different types of emitters under two management schemes of without and with chlorine injection into irrigation water. The assessment also included different orifice areas. It was observed that there was a reduction in emitter discharge induced by AP because of chlorine injection.

Clogging was evaluated by measuring through head loss across filters, and the filtration quality of different filters evaluated using different effluents.[87] It was observed that with the meat industry effluent, the poorest quality effluent, disk filters clogged more than other filter types. It was also found that the parameter that explained clogging, expressed as Boucher's filterability index, was different depending on the type of effluent and filter. They suggested that the best quality of filtration was achieved with a sand filter when the meat industry effluent was used. No significant differences were observed between the quality of filtration of disk and screen filters when operating with the secondary and tertiary effluents.

Hydraulic performance of three drip irrigation subunits were tested[88] using effluents: suspended solids and microorganism from Waste Water Treatment Plant of Castell-Platja d' Aro (Girona, Spain). All the subunits were operated intermittently for a total of 10 h per day, 5 days per week. The influence of different strategies of effluent treatment was evaluated on irrigation uniformity at the subunit level. It was concluded that with the secondary effluents, uniformity in subunit diminished considerably because of clogging of emitters. It was also observed that clogging occurred because of biological aspects.

The clogging mechanism of labyrinth channel in the emitter was examined,[89] using a three-dimensional numerical model of clogging analysis. Reynolds stress model with wall function was used to simulate the fluid flow in the Eulerian frame, and stochastic trajectory model was adopted to track the motion of the particles in a Lagrangian coordinate system without taking into account the agglomerating behavior of particles. The analytical results showed that in the labyrinth channel, low-velocity region developed ahead of each sawtooth and large vortex is shaped just behind it. Small particles were apt to deposit in those regions than those of large ones because of their better following behaviors. It was found that the potential clogging regions predicted by simulation were reasonably consistent with the experimental results. Further, it was also found that the particles ranging from 30 to 50 μm behaved best when passing through the labyrinth channel, and particle densities have a remarkable effect on the penetration only when their diameters were larger than 50 µm.

Root intrusion into emitters poses a threat to the long-term success of SDI systems, particularly in fibrous-rooted crops.[90] In this study, a Bermuda grass was grown in a greenhouse to examine the effectiveness of chemicals in preventing root intrusion into subsurface drip emitters in 2-year two-part experiments. During the first year of study, two acids (sulfuric and phosphoric) and two preemergence herbicides (trifluralin and thiazopyr) were tested on Bermuda grass grown in small pots. As an initial step for the emitter clogging experiment, the first-year experiment focused on the effectiveness of the chemicals in preventing overall root growth in pots saturated with either trifluralin or thiazopyr. It was found that only thiazopyr significantly inhibited root growth, and visual quality of shoot growth in the thiazopyr-treated pots was lower than the observed quality in the rest of the treatments and in nontreated Bermuda grass. During the second year, nine treatments were prepared based on the first-year study and were examined for the control of root intrusion into actual subsurface drip emitters. It was observed that emitters were completely free of roots with thiazopyr treatment at the highest concentration and with the trifluralin-impregnated emitter treatment under water stress. Authors concluded that root and rhizome growth was generally unaffected by treatment.

The effects of increasing sediment concentration in irrigation water and aperture size of screen filter were used on the sensitivity of some kinds of emitters to clogging.[91] The study included four concentrations of sediments in irrigation water (0, 70, 230, and 315 ppm) with aperture sizes of screen filter of 428.6, 179.3, 152, and 125 µm. The results indicated that the ratio of clogging differed from emitter to emitter under the same treatment because of the variations of emitter types and specifications. It was observed that the emitter LL (laminar, long path, type online) was the most sensitive to increase in suspended solids in irrigation water.

Experimental trials were conducted on the behavior of several kinds of filter and drip emitters using poor quality municipal wastewater.[22] The performance of the emitters and filters depended on the quality of the wastewater. It was suggested that total suspended solids (TSSs) influenced the percentage of totally clogged emitters, the mean discharge emitted, the emission uniformity, and the operating time of the filter between cleaning operations. Vortex emitters were more sensitive to clogging than labyrinth emitters, and no significant difference was observed between the same kind of emitter placed on soil or subsoil. Gravel media and disk filters assured better performance than screen filters. Finally, it was found that the use of wastewater with a TSS greater than 50 mg·L^{-1} did not permit optimal emission uniformity to be achieved.

Water from surface and underground sources was examined if it picked up particulate matter during conveyance of sands, silts, plant fragments, algae, diatoms, larvae, snails, fishes, etc.[92] It was found that as the flow slowed down and/or the chemical background of the water changes, chemical precipitates and/or microbial flocs and slimes began to form and grow, and thus micro irrigation emitter clogging occurred. It was suggested that section delineating the occurrences of chemical precipitated and the chemistry of acidification was employed to mitigate clogging caused by chemical precipitates. Finally, it was concluded that clogging resulting from the formation of microbial flocs and slimes was controllable by acidification as well as chlorination.

In a field experiment conducted at Hasanabad, Iran, by James,[19] five types of emitters with different nominal discharges, with or without self-flushing system and with or without pressure-compensating system, were evaluated under three management schemes: untreated well water (S_1), acidic treated water (S_2), and magnetic treated water (S_3) in order to reduce chemical clogging. Flow reduction rate, statistical uniformity coefficient (Uc), emission uniformity coefficient (Eu), and variation coefficient of emitters' performance in the field (Vf) were monitored. The emitter performance indices (Uc and Eu) decreased during the experiment because of emitter clogging. The Uc and Eu values in different management schemes confirmed that the acidification had better performance than the magnetic water in order to control emitter clogging and keep high distribution uniformity. Regarding Vf values, the priority of untreated and treated water was $S_2 > S_3 > S_1$ for each emitter.

An experiment was conducted to evaluate the effects of emitter clogging of four filtrations and six emitter types placed in laterals 87 m long using two different effluents with low suspended solid levels from a wastewater treatment plant for 1000 h.[93] It was found that only with the effluent that had a higher number of particles did the filter and the interaction of filter and emitter location have a significant effect.

A laboratory experiment was conducted to study the performance of three common emitter types with the application of freshwater and treated sewage effluent (TSE). The three types of emitters were the inline-labyrinth types of emitters with turbulent flow (E_1) and laminar flow (E_2) and online pressure-compensating type of emitter (E). It was found that for both freshwater and TSE treatment, the emitter clogging was more severe for emitter type E_2 because of its smaller flow path dimension and higher manufacturing coefficient of variation. Authors reported that main reason for emitter clogging was high pH and ion concentration in TSE treatments.[94]

The temporal variations of emitter discharge rate and the distribution of clogged emitters were studied in the drip irrigation system to quantify the impact of emitter clogging on system performance.[95] In the experiment, six types of emitters with or without pressure-compensating device and two types of water sources were considered. It was observed that different types of emitters had different susceptibilities to clogging. A more random distribution for clogged emitter was found to be suitable for sewage application. It was also reported that clogging of emitters deteriorated the system performance seriously.

The effects of three drip line flushing frequency treatments (no flushing, one flushing at the end of each irrigation period, and a monthly flushing during the irrigation period) were evaluated in surface and subsurface drip irrigation system operated using a wastewater treatment plant effluent for three irrigation periods of 540 h each.[96] It was found that drip line flow of the pressure-compensating emitter increased 8% over time, increased 25% in case of non-pressure-compensating emitter, and decreased 3% in subsurface drip lines by emitter clogging. It was concluded that emitter clogging was affected by the interaction between emitter location, emitter type, and flushing frequency treatment, and the number of completely clogged emitter was affected by the interaction between irrigation system and emitter type. The results of this study showed that the application of well saline water in drip irrigation system had the potential to induce emitter clogging. The concentration of Fe and Mg in well water was lower than the hazardous levels that could clog emitters. It was found that the flow rate reduction in emitters was affected by emitter characteristics and water treatment methods. Further, the acid injection treatment provided better performance than the magnetic field. On the other hand, less flow rate reduction occurred in emitters using acidic water.

A survey on the clogging level of emitters was conducted at some agricultural farms situated in Canakkale Turkey Onsekiz Mart University.[97] In the study, authors tested the emitters under pressures of 50, 100, 150, 200, 250, and 300 kPa in laboratory. It was found that some of the emitters were plugged on laterals used for 2–3 years in consequence of the tests. The laboratory test showed that 15.6% of 3-year used emitters collected from drip-irrigated land did not have any flow under an operating pressure of 100 kPa. Finally, it was suggested that drip irrigation system must be executed under prescribed pressure (100 kPa).

Laboratory and field testing were conducted in the hydraulic laboratory of the Agricultural Research Council at the Institute for Agricultural Engineering in South Africa.[98] Authors evaluated 3 dripper types and 10 dripper models from two dripper companies under controlled conditions. The field evaluation involved 42 surface drip systems throughout South Africa where different water quality conditions and management practices are present. Performance of these practices was evaluated in the field twice a year for two consecutive years, and after each evaluation, one dripper line was sampled and also tested in the laboratory. It was reported that the performance was affected by clogging because of water quality and lack of proper maintenance schedules.

2.2.5 ADSORPTION MECHANISM

A study was conducted on the adsorption of radiolabeled infectious poliovirus type-2 by 34 well-defined soils and mineral substrates.[33] Also these samples were analyzed in a synthetic freshwater medium containing 1 mM of $CaCl_2$ and 1.25 mM of $NaHCO_3$ at pH 7. It was found that in a model system, adsorption of poliovirus by Ottawa sand was rapid and reached equilibrium within 1 h at 4C. Near saturation, the adsorption was described by the Langmuir equation. The apparent surface saturation was 2.5×10^6 plaque-forming units of poliovirus per milligram of Ottawa sand. At low-surface coverage, adsorption was described by the Freundlich equation. It was observed that most of the substrates adsorbed more than 95% of the virus. Among the soils, muck and Genesee silt loam were the poorest adsorbents. Among the minerals, montmorillonite (MMT), glauconite, and bituminous shale were the least effective, and the most effective adsorbents were magnetite sand and hematite, which are predominantly oxides of iron. Correlation coefficients for substrate properties and virus adsorption revealed that the elemental composition of the adsorbents had little effect on poliovirus uptake. Substrate surface area and pH were not significantly correlated with poliovirus uptake. A strong negative correlation was found between poliovirus adsorption and both the contents of organic matter and the available negative surface charge on the substrates as determined by their capacities for adsorbing the cationic polyelectrolyte, poly diallyldimethylammonium chloride.

The adsorption processes were studied for the fabrication of layer-by-layer films using poly-o-methoxyaniline.[99] It was concluded that the amount of material adsorbed in any given layer depended on experimental parameters. It was observed that the H-bonding played a fundamental role in the adsorption of polyanilines on a glass substrate when the polymers were charged and electrostatic attraction was expected to predominate. The probability of adsorption increased in sites where some polymers were already adsorbed, which caused the roughness to increase with the number of layers. Also the electrostatic attraction was the predominant factor in the films.

Adsorption isotherm of Q-cresol from aqueous solution by granular activated carbon[100] was studied to investigate the equilibrium adsorption isotherms of Q-cresol from aqueous solution by a series of laboratory batch studies. A commercial norit granular activated carbon was used to evaluate the adsorption characteristic of Q-cresol at different temperatures of 30, 38, and 48°C. The effect of various initial concentrations (25–200 mg·L^{-1}) and time of adsorption on Q-cresol adsorption process were studied. The isotherm data using Langmuir and Freundlich isotherms were used to estimate the monolayer capacity values of activated carbon in the sorbate–sorbent system. The results revealed that the empirical Langmuir isotherm matched the observed data very well as compared with Freundlich isotherm. It was also found that the adsorption capacity of Q-cresol decreased with the increase in the adsorption temperature. The maximum adsorption capacity of 270 mg·g^{-1} was obtained by Q-cresol at a temperature of 30°C, 120 rpm, and 24 h of adsorption time.

In Sao Paulo State of Brazil,[101] the relationships were evaluated between sulfate adsorption and physical, electrochemical, and mineralogical properties of representative soils. The experimental results were subjected to variance, correlation, and regression analyses. When the adsorption was evaluated in the clay fraction, the kaolinite content was associated with low capacities of sulfate adsorption. However, no relationship was observed between the kaolinite soil content and the sulfate adsorption by the whole soil. No significant effects on sulfate adsorption were observed for individual hematite and goethite soil contents. On the other hand, the sum of hematite and goethite contents was related to sulfate adsorption.

Hussain et al.[102] studied phosphorous adsorption by five saline sodic soil samples collected from Faisalabad district. They prepared 0.01M of $CaCl_2$ solution with different concentrations of P and placed 3 g of soil in 30 mL solution of all P concentrations and kept the solution overnight and centrifuged, and the P in the supernatant solution was determined calorimetrically. They calculated the adsorption of P using the difference between the amount of P in supernatant and that added in solution and plotted the adsorption data according to Langmuir and Freundlich equation.

The adsorption behavior of Cu on three solid waste materials was investigated[103]: sea nodule residue, fly ash, and red mud. The effects of various parameters (pH of the feed solution, contact time, temperature, adsorbate and adsorbent concentrations, and particle size of the adsorbent) were studied for optimization of the process parameters. It was found that the adsorption of copper increased with increasing time, temperature, pH, and adsorbate concentration, and decreased with increasing initial copper concentration.

A laboratory experiment was conducted on synthesized hydrous stannic oxide (HSO) and Cr (VI) adsorption behavior by means of batch experiments,[104] to test the equilibrium adsorption data for the Langmuir, Freundlich, Temkin, and Redlich–Peterson equations. The scientists conducted batch adsorber tests by mechanical agitation (agitation speed: 120–130 rpm) using 0.2 g of HSO into a 100 mL polythene bottle with 50 mL of sorbate solution. Different concentrations of Cr solution were used in the range of 2.0–50.0 $mg \cdot L^{-1}$. They finally calculated the amount of adsorbed Cr by the difference of the initial and residual amounts in the solution divided by the weight of the adsorbent. It was concluded that the adsorption of Cr onto HSO took place by electrostatic interaction between adsorbent surface and species in the solution.

The phosphorus adsorption by Freundlich adsorption isotherm under rain-fed conditions was examined for 10 soil series of Pothwar Plateau.[105] These soils were treated with three different P fertilizers (diammonium phosphate (DAP), single superphosphate (SSP), and nitrophsophate (NP)) at equilibrium solution concentrations of 10, 20, 40, 60, and 80 $\mu g \cdot mL^{-1}$. Maximum Freundlich adsorption parameters (maximum adsorption ($\mu g \cdot g^{-1}$) and buffer capacity ($mL \cdot g^{-1}$)) were observed in Chakwal soil followed by Balkassar soil. The minimum values of these two Freundlich parameters were observed in Kahuta soil. It was observed that the maximum value of KOC in Chakwal soil with DAP, SSP and NP, while minimum value of KOC was observed in Bather soil with all fertilizers under investigation. A decrease in P adsorption with successive increase in equilibrium phosphorus solution concentration was recorded in all the soils under study.

A multiscale structure prediction technique was developed to study solution and adsorbed state ensembles of biomineralization proteins.[106] The algorithm, which employs a Metropolis Monte Carlo-plus minimization strategy, varies all torsional and rigid-body protein degrees of freedom. Authors applied this technique to fold statherin, starting from a fully extended peptide chain in solution, in the presence of hydroxyapatite (HAp) (001), (010), and (100) monoclinic crystals. Blind (unbiased) predictions capture experimentally observed macroscopic and high-resolution structural features and show minimal statherin structural change upon the adsorption. The dominant structural difference between solution and adsorbed states is an experimentally observed folding event in statherin's helical binding domain. Although predicted statherin conformers vary slightly at three different HAp crystal faces, geometric and chemical similarities of the surfaces allow structurally promiscuous binding. Finally, they compared blind predictions with those obtained from simulation biased to satisfy all previously published solid-state NMR (ssNMR) distance and angle measurements (acquired from HAp-adsorbed statherin). Atomic clashes in these structures suggested a plausible alternative interpretation of some ssNMR measurements as intermolecular rather than intramolecular. Finally, it was revealed that a combination of ssNMR and structure prediction could effectively determine high-resolution protein structures at biomineral interfaces.

An experiment was carried out on ion adsorption behavior of the polyacrylic acid–polyvinylidene fluoride-blended polymer.[107] Authors used polyvinylidene fluoride to remove copper from aqueous solutions. They prepared the polymer using thermally induced polymerization and phase inversion. The blended polymer was characterized by X-ray diffraction analysis, environmental scanning electron microscopy, X-ray photoelectron spectroscopy, and N_2 adsorption/desorption experiments. The sorption data was fitted to linearized adsorption isotherms of the Langmuir, Freundlich, and Dubinin–Radushkevich isotherm models. Further, they evaluated the batch sorption kinetics using pseudo-first-order, pseudo-second-order, and intraparticle diffusion kinetic reaction models. They found that ΔH° was greater than 0, ΔG° was lower than 0, and ΔS° was greater than 0, which showed that the adsorption of Cu (II) by the blended polymer was a spontaneous endothermic process. The adsorption isotherm fitted better to the Freundlich isotherm model, and the pseudo-second-order kinetics model gave a better fit to the batch sorption kinetics.

The zinc adsorption was studied in 10 soils of Punjab, varying in texture or calcareousness.[108] Authors executed the adsorption process by equilibrating 2.5 g soil in 25 mL of 0.01 M of $CaCl_2$ solution containing 0, 5, 10, 15, 20, 25, 70, and 120 mg of Zn per liter. Sorption data were fitted to Freundlich and Langmuir adsorption models. The data were best fitted in both linearized Freundlich and Langmuir equations as evidenced by higher correlation coefficient values ranging from 0.87 to 0.98. High clay contents ranging from 8 to 32% and $CaCO_3$ ranging from 4.46 to 10.6% promoted an increase in the amount of adsorbed zinc in these soils. They found that adsorption of Zn increased with the increasing level of Zn and also increased with increase in clay content, and $CaCO_3$ contents, and the maximum adsorption of Zn was observed in the Kotli soil, whereas the minimum was in the Shahdara soil series.

The experiment was undertaken on boron adsorption at the Institute of Soil and Environmental Sciences, University of Agriculture, Faisalabad, Pakistan, in five different textured calcareous soils of Punjab.[109] They executed the adsorption process by equilibrating 2.5 g of soil in 0.01 M of $CaCl_2$ solution containing different concentrations of boric acid for 24 h. They estimated the boron adsorption using Langmuir and Freundlich models. They concluded that the Freundlich model was better than the Langmuir model.

The iodide adsorption was evaluated to compare the sorption behavior of iodate and iodide.[110] They collected typical soil samples at 17 locations across China. Batch experiments of iodate and iodide adsorption were carried out by shaking soil samples equivalent to 2.5 g dry weight with 25 mL of iodine (either iodate or iodide) solution. This was performed in centrifuge tubes fitted with caps, on an end-over-end shaker (160 rpm) at 25°C and shaken for 40 h. For the sorption isotherm studies, concentrations of KIO_3 in the solution were 0, 1, 2, 4, 6, 8 mg·L^{-1} for the two soil types from Xinjiang Province and Beijing City. The results indicated that the capacity of iodate adsorption by the five soils was markedly greater than that of iodide. Furthermore, detailed comparison of sorption parameters based on the Langmuir and Freundlich adsorption equations supported this finding showing a greater adsorption capacity for iodate than for iodide due to higher k_2 values of iodate than those of iodide.

Adsorption studies of zinc and copper ions were attempted on MMT. The adsorption mechanism of the metal adsorptions was studied by the measurement of UV–VIS DRS of Zn MMT and Cu MMT.[111] They used zinc nitrate, copper sulfate, ammonium chloride, and ethylenediamine (EDA). For adsorption procedure, they prepared the EDA–MMT by shaking MMT in concentrated EDA (0.9 g·cm^{-3}) for 24 h, and then filtered and dried at 105°C for 2 h. They saturated the MMT and EDA–MMT metals by shaking in the solutions of 5 mmol·L^{-1} of Cu^{2+} and 20 mmol·L^{-1} of Zn^{2+} at 170 rpm for 24 h. Then they centrifuged the suspension for 24 h and analyzed the filters for zinc and copper using atomic absorption spectrometry. They concluded that using the DRIFT method, the amount of interlayer water in Zn MMT and Cu MMT was similar.

Research was conducted on bottlebrush polymers and their adsorption on surfaces and their interactions.[112] By small-angle scattering techniques, they studied the solution conformation and interactions in solution. Surfactant binding isotherm measurements, NMR, surface tension measurements, as well as SAXS, SANS, and light scattering techniques were utilized for understanding the association behavior in bulk solutions. The adsorption of the bottlebrush polymers onto oppositely charged surfaces was explored using a battery of techniques: reflectometry, ellipsometry, quartz crystal microbalance, and neutron reflectivity. The combination of these techniques allowed the determination of adsorbed mass, layer thickness, water content, and structural changes occurring during layer formation. The adsorption onto mica was found to be very different from that on silica, and an explanation for this was sought by employing a lattice mean-field theory. The model was able to reproduce a number of salient experimental features characterizing the adsorption of the bottlebrush polymers over a wide range of compositions, spanning from uncharged bottlebrushes to linear polyelectrolyte. The interactions between bottlebrush polymers and anionic surfactants in adsorbed layers were elucidated using ellipsometry, neutron reflectivity, and surface force measurements.

The adsorption characteristics of phosphorus were evaluated onto soil with the Langmuir, Freundlich, and Redlich–Peterson isotherms by both the linear and nonlinear regression methods.[113] The adsorption experiment was conducted at temperatures of 283, 288, 298, and 308°K to choose the appropriate method and obtain the credible adsorption parameters for soil adsorption equilibrium studies. The results showed that the nonlinear regression method was a better choice to compare the better fit of isotherms for the adsorption of phosphorus onto laterite. Both the two-parameter Freundlich and the three-parameter Redlich–Peterson isotherms had higher coefficients of determination for the adsorption of phosphorus onto laterite at various temperatures.

A study on nZVI particles was conducted to investigate the removal of Cd_2^+ in the concentration range of 25–450 $mg \cdot L^{-1}$.[114] The effect of temperature on kinetics and equilibrium of cadmium sorption on nZVI particles was thoroughly examined. They found that the maximum adsorption capacity of nZVI for Cd^{2+} was 769.2 $mg \cdot g^{-1}$ at 297 K. Thermodynamic parameters were change in the free energy (G_0), the enthalpy (H_0), and the entropy (S_0). These results suggested that nZVI can be employed as an efficient adsorbent for the removal of cadmium from contaminated water sources.

A study was conducted on the adsorption behavior of Mn from an agricultural fungicide in two southwestern Nigeria soils using batch equilibrium test.[115] They applied two mathematical models described by Langmuir's and Freundlich's adsorption equations. From the isotherm analysis, they found that the sorption of Mn to the two soil types considered was best described by Freundlich model, and the maximum adsorption capacities (kf) obtained from this model were 96.64 $g \cdot mL^{-1}$ and 30.76 $g \cdot mL^{-1}$ for Egbeda and Apomu soils, respectively. These maximum adsorption capacities occurred at a solution pH of 5 for both soils. Finally, they found that the solution pH values of 3 and 4 were not significantly different as well as the solution pH values of 5 and 6 in their effects on the amount of Mn adsorbed.

A laboratory study was conducted at the research and development (chemical laboratory) of product development in M/s Jain Irrigation System, Plastic Park, Jalgaon, Maharashtra, India, during 2010–2011 to know the adsorption mechanism using adsorption characteristics of Langmuir and Freundlich equations for granules of PVC, LLDPE, silicone diaphragm (rubber button), and HDPE. The results indicated that the percentage of clogging was maximum for LLDPE indicating 6.94–11.1% from day 1 to day 15 compared with other grinded materials. The minimum percentage of clogging was recorded as 1.12–3.03% for PVC. The results demonstrated that

LLDPE is more susceptible for clogging in drip irrigation system.[116,117] For clogging test, two different types of emitter type A (2 lph) and B (4l lph) were used. The results of clogging test demonstrated that emitter type A was more susceptible for clogging compared with type B. The emitter type B was superior compared with emitter type A.

2.3 EMITTER CLOGGING: A MENACE TO DRIP IRRIGATION

The clogging agents are summarized in Table 2.2. The clogging problems can be classified as minor, moderate, and severe (Table 2.3). The physical contributors include mineral particles of sand, silt, and clay, and debris that are too large to pass through the small openings of filters and emission devices. Silt and clay particles that are usually much smaller than the smallest passages are often deposited in the low-velocity areas of the laterals where they coagulate to form masses large enough to clog emission devices. Coating of clay particles in filters and emission devices can also reduce water flow.

TABLE 2.2   Water Quality Factors Affecting the Clogging of Drip Irrigation Systems

Physical (Suspended Solids)	Chemical (Precipitation)	Chemical (Bacteria and Algae)
Inorganic particles: Sand, silt, clay	Calcium or magnesium carbonates	Filaments
Plastic	Calcium sulfate	Slime
Organic particles	*Heavy metals*	*Microbial decomposition*
Aquatic plants (phytoplankton/algae)	Oxides, hydroxides	Iron
Aquatic animals (zooplankton)	Carbonates	Sulfur
Bacteria	Silicates and sulfides	Manganese
	Oils or other lubricants, fertilizers	
	Phosphate Aqueous ammonia Iron, copper, zinc, manganese	

TABLE 2.3 *(Continued)*

TABLE 2.3 Severity of Clogging Hazard

Types of Agents	Clogging Hazard		
	Minor	**Moderate**	**Severe**
Physical			
Suspended solids	<50	50–100	>100
Chemical			
pH	<7.0	7.0–8.0	>8.0
Dissolved solids[a]	<500	500–2000	>2000
Manganese[a]	<0.1	0.1–1.5	>1–5
Total iron[a]	<0.2	0.2–1.5	>1.5
Hydrogen sulfide[a]	<0.2	0.2–2.0	>2.0
Biological			
Bacterial population[b]	<10,000	10,000–50,000	>50,000

[a]Maximum measured concentration from a representative number of water samples using standard procedures for analysis ($mg·L^{-1}$).
[b]Maximum number of bacteria per milliliter can be obtained from portable field samplers and laboratory analysis. Bacteria populations do reflect increased algae and microbial nutrients.

Salinity is an important water quality factor in irrigation and does not contribute to emitter clogging unless the dissolved ions interact with each other to form precipitates or promote slime growth. When irrigation water contains soluble salts, crusts of salt often form on emission devices as the water evaporates between irrigations. If the salt does not dissolve during the subsequent irrigation, crust accumulation will continue, and clogging of the emission device will usually result. The factors conducive to chemical precipitation are high concentrations of calcium and magnesium and bicarbonate ions and relatively high pH of water. Temperature is also a factor because the solubility of calcium carbonate precipitates decrease with an increase in temperature.

The corrective treatment for controlling clogging depends on the types of clogging agents. Many agents can be removed using settling basins, water filtration, and/or periodic flushing of filters, mainlines, laterals, and emission devices. Injection of acids, algicides, and bactericides is a common treatment used to control chemically and biologically caused clogging.

By performing certain water analyses, possible problems can be estimated. This is especially advisable before a new drip system is installed. The factors are rated in terms of an arbitrary clogging hazard, ranging from minor to severe, and are presented in Table 2.3. Following precautionary measures are mostly related to the physical and chemical properties of water:

Clogging Factors	Remedial Measures
Poor physical quality water	Filtration
Poor chemical quality water	Flushing

2.3.1 FILTRATION

Adequate filtration requires the processing of water entering the system. The particle size, which can be tolerated in the system, depends on the emitter construction. Typically, the recommendation is for the removal of particles larger than 1/10th the diameter of the orifice or the flow passage of the emitter. One reason for this is that several particles may group together and obstruct the passageway. This is typical with organic particles having about the same density as water. Another reason is that heavier inorganic particles (fine and very fine sands) tend to settle and deposit in slow-flow zones, particularly inside walls of laminar flow emitters where the flow rate is slow. The result of clogging may not be rapid, but it is inevitable. It may be necessary to use a 200-mesh screen, which has a 0.074 mm (0.0029 inch) hole size even with a passageway of 1 mm (0.04 inch) in cross section. Most manufacturers recommend removing particles larger than 0.075 mm or 0.15 mm (0.003 or 0.006 inch), but some allow particles as large as 0.6 mm (0.024 inch). Table 2.4 summarizes the minimum-sized particles that can be removed by several of these devices. Removal of suspended particles is usually required to get optimum performance of the drip system, because irrigation water is rarely free of suspended material. Settling basins, sand/media filters, screens, cartridge filters, disk filters, and centrifugal separators are the primary devices used to remove suspended material. The best way to reduce or prevent clogging is by adequate filtration.

1. Suspended organic matter and clay particles may be separated with gravel filters or screen filters.
2. Filter cleaning becomes necessary if pressure drops significantly between the entry and the exit sides of the filter. As a rule, it is customary to clean the filters when the allowable pressure drops (about 4 m or 0.4 kg·cm^{-2} or 6 psi).

TABLE 2.4 Filter Effectiveness

Filter Type	Size Range (μm)
Sand media	5–100
Sand media	>20
Screen	>75
Screen (100–200 mesh)	75–150
Screen (200 mesh)	>100

TABLE 2.4 *(Continued)*

Filter Type	Size Range (μm)
Sediment basins	>40
Separator[a]	>74
Separator[a] (two stage)	>44
Slotted cartridge	>152

[a]Separators remove 98% of particles larger than the size indicated.

2.3.2 SETTLING BASIN

A settling basin, settling pond, or decant pond is an earthen or concrete structure using sedimentation to remove settleable matter and turbidity from wastewater. The basins are used to control water pollution in diverse industries such as agriculture, aquaculture, and mining. Turbidity is an optical property of water caused by the scattering of light by material suspended in that water. Although turbidity often varies directly with weight or volumetric measurements of settleable matter, correlation is complicated by variations in size, shape, refractive index, and specific gravity of suspended matter. Settling ponds may be ineffective at reducing turbidity caused by small particles having specific gravity low enough to be suspended by Brownian motion.

Settling basins are designed to retain water long enough so that suspended solids can settle to obtain a high-purity water in the outlet and also provide the opportunity for pH adjustment. Thickeners, clarifiers, and hydrocyclones, as well as membrane filtration are also used in the field. Compared with those processes, settling basins have a simpler and cheaper design, and fewer moving parts, demanding less maintenance, despite requiring cleaning and vacuuming of the quiescent zones at least once every 2 weeks. Also there can be more than one settling basins in series.

Settling basins or reservoirs (Fig. 2.2) can remove large volumes of sand and silt. The minimum size of a particle that can be removed depends on the time that sediment-laden water is detained in the basin. Longer detention times are needed to remove smaller particles. The basin should be constructed so that water entering the basin takes at least a quarter of an hour to travel to the system intake. In this length of time, most inorganic particles larger than 80 μm (about 200 mesh) will settle. A basin 1.2 m deep × 3.3 m wide × 13.7 m long (4 × 10 × 45 ft) is required to provide a quarter hour retention time for a 57 lps (900 gpm) stream. Removal of clay-sized particles requires several days and is not practical unless flocculating agents such as alum and/or polyelectrolytes are used. Settling basins may need to be cleaned several times a year when large quantities of water with high concentrations of sediment are being passed through the basin.

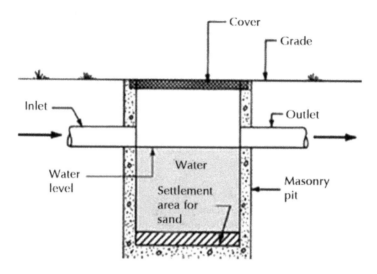

FIGURE 2.2   Setting basin to intercept sand particles.

Algae growth and wind-blown contaminants can be severe problems in settling basins. The sides and bottom of the basin should be lined to discourage vegetative growth, and chemical treatment with chlorine or copper sulfate may be required to control algae. Because of these problems, settling basins are recommended for use with drip systems only in extreme circumstances.

2.3.3   SAND MEDIA FILTERS

Sand or gravel or media filters (Fig. 2.3) are used for filtering out heavy loads of very fine sands and organic materials. These consist of layered beds of graduated sand and gravel placed inside one or more pressurized tanks. They effectively remove suspended sands, organic minerals, and most other suspended substances from the surface and groundwater. Also, long, narrow particles, such as some algae or diatoms, can be caught in the multilayered sand bed than on the surface of a screen. They do not remove very fine particles (i.e., silt and clay) or bacteria. Media filters are relatively inexpensive and easy to operate. It is generally recommended that the filter material be as coarse textured as possible but fine enough to retain all particles larger than one-sixth the size of the smallest passageway in the drip irrigation system. Filter materials should be large enough not to be removed during filter cleaning processes. A recommended practice is to use a screen filter downstream from the media filter to pick particles that escape during backwashing.

FIGURE 2.3 A typical sand media filter and backwash process for cleaning (bottom figure).

Factors that affect filter characteristics and performance are water quality, type and size of sand media, flow rate, and allowable pressure drop. A sand media filter can handle larger loads of contaminants than a screen of comparable fineness. It can do it with less frequent back flushing and a smaller pressure drop.

However, sand filters are considerably more expensive. They are generally used only when a screen filter would require very frequent cleaning and attention or to remove particles smaller than 0.075 mm (0.003 inch).

The sand media used in most drip irrigation filters is designated by numbers. Numbers 8 and 11 are crushed granite, and numbers 16, 20, and 30 are silica sands. The mean granule size in micrometers for each media number is approximately 1900, 1000, 825, 550, and 349 for numbers 8, 11, 16, 20, and 30, respectively.

At a flow velocity of 17 $L \cdot sec^{-1} \cdot m^{-2}$ (25 $gpm \cdot ft^{-2}$) of bed, numbers 8 and 11 crushed granite remove most particles larger than 1/12th of the mean granule size or approximately 160 and 80 µm, respectively. The silica sand numbers 16, 20, and 30 remove particles approximately 1/15th the mean granule size or approximately 60, 40, and 20 µm, respectively.

Typically, the initial pressure drop across numbers 8, 10, and 16 media is between 14 and 21 kPa (2 and 3 psi). For number 20 and 30 media, it is approximately 34 kPa (5 psi). The rate of pressure drop increase is usually linear with time for a given quality of water and flow rate. Assuming 1.0 unit of pressure drop per unit of time for a number 11 media, the units of pressure drop per unit in time across the other media would be 0.2 unit for number 8 media, 2 units for number 16 media, 8 units for number 20 media, and 15 units for number 30 media. For example, if it takes 24 h for the pressure drop to increase by 34 kPa (5 psi) across a number 11 media, it would take only about 3 h for the same increase across a number 20 media. The maximum recommended pressure drop across a sand filter is generally about 70 kPa (10 psi or 0.6 $kg \cdot cm^{-2}$).

2.4 SUMMARY

The orifices in the drip lines or the emitters emit water to the soil. The emitters allow only the discharge of few liters or gallons per hour. The emitters have small orifices, and these can be easily obstructed. For a trouble-free operation, one should follow these considerations: pay strict attention to filtration and flushing operation. Maintain an adequate operating pressure in the main, sub main, and lateral lines. Flushing and periodic inspection of the drip irrigation system are a must. For effective filtration efficiency, we must maintain the system in good condition, and it is not obstructed by the clogging agents. For this, pressure gauges are installed at the entrance and the exit of a filter. The frequency of flushing depends on the water quality. Some recommendations for an adequate maintenance are cleaning with pressurized air, acids, and chlorine.

The review of literature on studies on clogging mechanism in drip irrigation system indicated that the clogging of emitters can occur because of three clogging agents[20]: (a) physical clogging, (b) chemical clogging, and (c) biological clogging. Most of the clogging studies suggested flushing, chlorination, combination of filtration, emitter design, and field practices in order to reduce emitter clogging. This is all about the temporary solution to reduce clogging. But the information is not available on the initial adsorption mechanism of clogging of emitters in drip irrigation system for evolving strategies for reducing clogging, and such studies were very few or not done previously.

KEYWORDS

- acidification
- adsorption
- Brazil
- China
- chlorination
- clogging
- clogging mechanism
- clogging, biological
- clogging, chemical
- clogging, physical
- drip irrigation
- effluent
- emitter
- flushing
- Freundlich equation
- India
- Langmuir equation
- Nigeria
- Pakistan
- polymers
- saline water
- solid waste
- USA
- wastewater
- water shortage

REFERENCES

1. Iyer, R. R. *Water: Perspectives, Issues, Concerns*; Sage Publications: New Delhi, 2003.
2. MoWR. Report of the Working Group on Water Availability for Use, National Commission for Integrated Water Resources Development Plan; Ministry of Water Resources, Government of India: New Delhi, 1999.
3. Saleth, R. *Water Institutions in India: Economics, Law and Policy*; Commonwealth Publishers: New Delhi, 1996.
4. Evans, R. G. *Micro Irrigation*; Washington State University, Irrigated Agriculture Research and Extension Center: Prosser, WA, USA, 2000.

5. Powell, N. L.; Wright, F. S. Subsurface micro irrigated corn and peanut: effect on soil pH. *J. Agric. Water Manag.* 1998, 36, 169–180.

6. Sahin, U.; Anupali, O.; Döumez, M.; Sahin F. Biological treatment of clogged emitters in a drip irrigation system. *J. Environ. Manage.* 2005, 76(4), 301–479.

7. International Commission on Irrigation and Drainage (ICID). *Sprinkler and Micro-Irrigated Areas in Some ICID Member Countries.* 2000. http://www.icid.org/ index_e.html.

8. INCID. *Indian National Committee on Irrigation and Drainage.* 2008. http://www.icid. org/v_india.pdf.

9. IARI.2011.http://www.iari.res.in/index.php?option=com_jumi&fileid=24&Itemid=664.

10. Bresler, E. *Trickle-Drip Irrigation: Principles and Applications to Soil-Water Management.* In *Advances in Agronomy*; Brady, N. C., Ed.; Academic Press: New York, 1975; Vol. 29; p 343–393.

11. Nakayama, F. S. *Trickle Irrigation for Crop Production*; Elsevier Science Publishers: Amsterdam, Netherlands, 1986; p 383.

12. Ahmed, B. A. O.; Yamamoto, T.; Fujiyama, H.; Miyamoto, K. Assessment of emitter discharge in micro irrigation system as affected by polluted water. *Irrig. Drain. Syst.* 2007, 21, 97–107.

13. Chigerwe, J.; Manjengwa, N.; Van der Zaag, P. Low head drip irrigation kits and treadle pumps for smallholder farmers in Zimbabwe: a technical evaluation based on laboratory tests. *Phys. Chem. Earth*, 2004, 29, 1049–1059.

14. De Kreij, C.; Van der Burg, A. M. M.; Runia, W. T. Drip irrigation emitter clogging in Dutch greenhouses as affected by methane and organic acids. *Agric. Water Manage.* 2003, 60, 73–85.

15. Povoa, A. F.; Hills, D. J. Sensitivity of micro irrigation system pressure to emitter plugging and lateral line perforations. *Trans. Am. Soc. Agric. Eng.* 1994, 37(3), 793–799.

16. Gilbert, R. G.; Nakayama, F. S.; Bucks, D. A.; French, O. F.; Adamson, K. C. Trickle irrigation: emitter clogging and flow problems. *J. Agric. Water Manage.* 1981, 3, 159–178.

17. Pitts, D. J.; Haman, D. Z.; Smajstrla, A. G. *Causes and Prevention of Emitter Plugging in Micro Irrigation Systems*; University of Florida, Cooperative Extension Service Bulletin: Florida, 1990; Vol. 258.

18. Gilbert, R. G.; Ford, H. W. *Operational Principles/Emitter Clogging.* In *Trickle Irrigation for Crop Production*; Nakayama, F. S., Bucks, D. A. Ed.; Elsevier: Amsterdam, 1986; p 142–163.

19. James, L. G. *Principles of Farm Irrigation System Design*; John Wiley: New York, 1988; p 287–297.

20. Goyal Megh R. (Senior Editor-in-Chief). *Book Series on Research Advances in Sustainable Micro Irrigation*; Apple Academic Press Inc.: Oakville, ON, Canada, 2015; Vol. 1–10.

21. Nakayama, F. S.; Bucks, D. A. Water treatments in trickle irrigation systems. *J. Irrig. Drain. Div. Am. Soc. Civil Eng.* 1991, 104S(IR1), 23–34.

22. Capra, A.; Scicolone, B. Recycling of poor quality urban wastewater by drip irrigation systems. *J. Clean. Prod.* 2006, 15, 1529–1534.

23. Ould, A.; Yamamoto T.; Fujiyama H.; Miyamoto K. Assessment of emitter discharge in micro irrigation system as affected by polluted water. *J. Irrig. Drain.* 2007, 21, 97–107.

24. English S. D. Filtration and Water Treatment for Micro-Irrigation. In Drip/Trickle Irrigation in Action, Proceeding of the third International Drip Irrigation Congress, November 18-21; Fresno, California, 1985; Vol 1; p 54–57.

25. Merriam, J. L.; Keller, J. *Farm Irrigation System Evaluation: A Guide for Management*; Agricultural Irrigation Eng. Dept. Utah State University Logan: Utah, 1978; p 271
26. Water Global Researcher. 2008. http://www.globalresearcher.com.
27. Harris, G. *Sub Surface Drip Irrigation System Components*; Queen Land Government, Department of Primary Industries and Fisheries: Australia, 2005.
28. Hills, D. J.; Brenes, M. J. Micro irrigation of wastewater effluent using drip tape. *J. Appl. Agric. Eng.* 2001, 17(3), 303–308.
29. Yingduo, Y.; Gong, S.; Di, X.; Wang, J.; Xiaopeng, M. Effects of Treflan injection on winter wheat growth and root clogging of subsurface drippers. *Agric. Water Manage.* **2010,** 97(5), 723–730.
30. ASAE. Design and Installation of Micro irrigation Systems; ASAE EP405: St. Joseph, MI, USA, 2001.
31. Howell, T. A.; Stevenson, D. S.; Aljibury, F. K.; Gitlin, H. M.; Wu, I. P.; Warrick, A. W.; Raats, P. A. C. *Design and Operation of Trickle (Drip) Systems*. In *Design and Operation of Farm Irrigation Systems*; Jensen, M. E. Ed., ASAE: St Joseph, MI, 1983, *Monograph by American Society of Agriculture Engineers.*
32. Hills, D. J.; Tajrishy, M. A.; Tchobanoglous, G. The influence of filtration on ultraviolet disinfection of secondary effluent for micro irrigation. *Trans. Am. Soc. Agric. Eng.* 2000, 43(6), 1499–1505.
33. Moore, S.; Dene, H. T.; Lawrence, S. S.; Michael, M. R.; Fuhs G. W. *J. Appl. Environ. Microbiol,* 1981 42(6), 963–975.
34. Al-Amound, A.; Saeed, M. The Effect of Pulsed Drip Irrigation on Water Management. In Proceeding the 4th International Micro irrigation Congress, 1988; 4b-2.
35. Stephen, D. E. Filtration and Water Treatment for Micro irrigation. Drip/ Trickle Irrig. in Action. In The third International Drip /Trickle Irrigation Congress; ASAE Publications, 1985; Vol. 1.
36. Ford, H. W.; Tucker, D. P. H. Water quality measurement for drip irrigation systems. *J. Fla. Agric. Exp. Stn.* 1974, 5598, 58–60.
37. Parchomchuk, P. Temperature effect on emitter discharge rates. *Trans. Am. Soc. Agric. Eng.* 1976, 19(4), 690–692.
38. Gilbert, R. G.; Nakayama, F. S.; Bucks, D. A.; French O. F. Trickle irrigation: emitter clogging and other flow problems. *J. Agric. Water Manage.* 1980, 3, 159–177.
39. Gilbert, R. G.; Nakayama, F. S.; Bucks, D. A.; French, O. F.; Adamson, K. C.; Johnson, R. M. Trickle irrigation: predominant bacteria in treated Colorado River water and biologically clogged emitters. *J. Irrig. Sci.* 1982, 2(3), 123–132.
40. Adin, A. Clogging in irrigation systems reusing pond effluents and its prevention. *Water Sci. Technol.* 1987, 19(12), 323–328.
41. Feng, F.; Li, Y.; Guo, Z.; Li, J.; Li, W. Clogging of emitter in subsurface drip irrigation system. *Trans. CSAE,* 2004, 20(1), 80–83.
42. Assuncao, T. P. R.; Jose E. S. P.; Christiane, C. Chemical treatment to unclog dripper irrigation systems due to biological problems. *J. Sci. Agric. (Piracicaba, Braz.),* 2008, 65(1), 1–9.
43. Wei, Q.; Lu, G.; Liu, J.; Shi, Y.; Dong, W.; Huang, S. Evaluations of emitter clogging in drip irrigation by two-phase flow simulations and laboratory experiments. *J. Comput. Electron. Agric.* 2008, 63(2), 294–303.

44. Padmakumari, O.; Sivanappan, R. K. Study on Clogging of Emitters in Drip Irrigation System. In Paper Presented at the 1985 Proceedings of the 3rd International Drip Irrigation System Congress; 1985; p 80–83.

45. Meyer, J. L.; Snyder, M. J.; Valenzuela, L. H.; Strohman, R. Liquid polymers keep drip irrigation lines from clogging. *J. Calif. Agric.* 1998, 45(1), 116–120.

46. Rav-Acha, C.; Kummel, M.; Salamon, I.; Adin, A. The effect of chemical oxidants on effluent constituents for drip irrigation. *Science,* **2000**, 29(1), 119–129.

47. Enciso, J.; Porter, D. *Maintaining Subsurface Drip Irrigation System*; Texas Cooperative Extension Service, The Texas A & M University, 2001, L-5401, 10-01.

48. Anonymus. *Micro Irrigation*; Jain Irrigation Systems Ltd, 2002.

49. Granberry, D. M.; Harrison, K. A. *Drip Chemigation; Injecting Fertilizer, Acid and Chlorine*; University of Georgia College of Agricultural and Environmental Science, US Department of Agriculture, 2005; p. 55.

50. Aali, K. A.; Liaghat, A. The effect of acidification and magnetic field on emitter clogging under saline water application. *J. Agric. Sci.* 2009, 1(1), 112–117.

51. Netafim. *Drip System Operation and Maintenance*; Netafim Irrigation USA, 2009. http://www.netafimusa.com.

52. Li, J.; Chen, L. Assessing emitter clogging in drip irrigation system with sewage effluent. *Trans. Am. Soc. Agric. Biol. Eng.* 2009. http://*www.asabe.org.*

53. Ford, H. W.; Tucker, D. P. H. Blockage of drip irrigation filters and emitters by iron – sulfur – bacterial products. *J. Hortic. Sci.* 1975, 10(1), 62–64.

54. Ford, H. W. Controlling slimes of sulfur bacteria in drip irrigation systems. *J. Hortic. Sci.* 1976, 11, 133–135.

55. Boman, B. J. Effects of orifice size on micro sprinkler clogging rates. *J. Appl. Agric. Eng.* 1995, 11(6), 839–843.

56. Stansly, P. A.; Pitts, D. J. Pest damage to micro-irrigation tubing: causes and prevention. *Proc. Fla. State Hortic. Soc.* 1990, 103, 137–139.

57. Childers, C. C.; Futch, S. H.; Stange, L. A. Insect (Neuroptera: Lepidoptera) clogging of a micro sprinkler irrigation system in Florida citrus. *Fla. Entomol.* 1992, 75(4), 601–604.

58. Brushwein, J. R.; Matthews, C. H.; Childers, C. C. *Selenisa sueroides* (Lepidoptera: Noctuidae): a pest of sub-canopy irrigation systems in citrus in Southwest Florida. *Fla. Entomol.* 1989, 72(3), 511–518.

59. Boman, B. J.; Bullock, R. C. Damage to micro sprinkler riser assemblies from *Selenisa sueroides* caterpillars. *J. Appl. Eng. Agric.* 1994, 10(2), 221–223.

60. Boman, B. J. Insect Problems in Florida Citrus Micro Irrigation Systems. In Proc. 4th National Irrigation Symposium; Evans, R. G., Benhan, B. L., Trooien, T. P., Ed.; American Society of Agricultural Engineers: St. Joseph, Michigan, 2000; p 409–415.

61. Ruskin, R.; Ferguson, K. R. Protection of Subsurface Drip Irrigation Systems from Root Intrusion. In Proceedings of Irrigation Association 19th Annual Meeting; San Diego, CA, 1998; p 1–3.

62. Costerton, J. W.; Stewart, P. S. Battling Biofilms. *Sci. Am.* 2001, 285(1), 75–81.

63. Bucks, D. A.; Nakayama, F. S.; Gilbert, R. G. Trickle irrigation water quality and preventive maintenance. *Agric. Water Manage.* 1979, 2, 149–162.

64. Adin, A.; Sacks, M.; Dripper clogging factors in wastewater irrigation. *J. Irrig. Drain. Eng.* 1991, 117(6), 813–825.

65. Yavuz, K. D.; Okan, E.; Erdem, B.; Merve, D. Emitter clogging and effects on drip irrigation systems performances. *Afr. J. Agric. Res.* 2010, 5(7), 532–538.

66. O'Toole, G. A.; Kolter, R. Flagellar and twitching motility are necessary for *Pseudomonas aeruginosa* biofilm development. *J. Mol. Microbiol.* 1998, 30(2), 295–304.

67. Davey, M. E.; O'Toole, G. A. Microbial biofilms: from ecology to molecular genetics. *Microbiol. J. Mol. Biol.* 2000, 64(4), 847–867.

68. Davies, D. G.; Parsek, M. R.; Pearson, J. P.; Iglewski, B. H.; Costerton, J. W.; Greenberg, E. P. The involvement of cell-to-cell signals in the development of bacterial biofilm. *Science*, 1998, 280(5361), 295–298.

69. Ravina, E.; Paz, Z.; Sofer, A.; Marcu, A.; Sagi, S. G. Control of emitter clogging in drip irrigation with reclaimed wastewater. *J. Irrig. Sci.* 1992, 129–139.

70. Sagi, G.; Paz, E.; Ravina, I.; Marcu, S. A.; Yechiely, Z. Clogging of Drip Irrigation Systems by Colonial Protozoa and Sulfur Bacteria. In Proceedings of 5th International Micro irrigation Congress; Orlando FL, 1995.

71. Ravina, E. P.; Sofer, Z.; Marcu, A.; Shisha, A.; Sagi, G.; Lev, Y. Control of clogging in drip irrigation with stored treated municipal sewage effluent. *Agric. Water Manage.* 1997, 33(2–3), 127–137.

72. Nakshabandi, G. A.; Saqqar, M. M.; Shatanawi, M. R.; Fayyad, M.; Al-Horani, H. Some environmental problems associated with the use of treated wastewater for irrigation in Jordan. *Agric. Water Manage.* 1997, 34, 81–94.

73. Coelho, R. D.; Resende, R. S. *Biological Clogging of Netafim's Drippers and Recovering Process Through Chlorination Impact Treatment*; ASAE Paper Number: 012231; ASAE: Sacramento, CA, USA, 2001.

74. Trooien, T. P.; Lamm, F. R.; Stone, L. R.; Alam, M.; Clark, G. A.; Rogers, D. H.; Clark, G. A.; Schlegel, A. J. *Subsurface Drip Irrigation Using Livestock Wastewater*. 2009. http://www.ksre.ksu.edu/sdi.

75. Wu, F.; Fan, Y.; Li, H.; Guo, Z.; Li, J.; Li, W. Clogging of emitter in subsurface drip irrigation system. *Trans. China State Agric. Eng.* 2004, 20(1), 80–83.

76. Oron, G.; Shelef, G.; Truzynski, B. Trickle irrigation using treated wastewaters. *J. Irrig. Drain.* 1979, 105(IR2), 175–186.

77. Nakayama, F. S.; Bucks, D. A.; French, O. F.; Adamson, L. Trickle irrigation: emitter clogging and other flow problems. *Agric. Water Manage.* 1981, 3, 159–178.

78. Smajstrla, A. G.; Koo, R. C. J.; Weldon, J. H.; Harrison, D. S.; Zazueta, F. S. Clogging of trickle irrigation emitters under field conditions. *Proc. Fla. State Hortic. Soc.* 1983, 96, 13–17.

79. Schwankl, L. J.; Prichard, T. L. Clogging of buried drip irrigation system. *J. Calif. Agric.* 1990, 44(1).

80. Boman, B. J. Clogging characteristics of various micro sprinkler designs in a mature citrus grove. *Proc. Fla. State Hortic. Soc.* 1990, 103, 327–330.

81. Tajrishy, M.; Hills, D.; Tchobanoglous, G. Pretreatment of secondary effluent for drip irrigation. *J. Irrig. Drain.* 1994, 120(4), 716–731.

82. Alam, M.; Rogers, D. H. *Filteration and Maintenance Considerations for Subsurface Drip Irrigation Systems*; Kansas State University, Agricultural Experiment Station and Cooperative Extension Service: Manhattan, Kansas, 2002.

83. Ribeiro, T. A. P.; Paterniani, J. E. S.; Coletti, C. Chemical treatment to unclogg dripper irrigation systems due to biological problems. *J. Agric. Sci. (Piracicaba, Braz)*, 2008, 65(1), 1–9.

84. Taylor, H. D.; Bastos, R. K. X.; Pearson, H. W.; Maro, D. D. Drip irrigation with waste water stabilization pond effluents solving the problem of emitter fouling. *J. Water Sci. Technol.* 1995, 31(12), 417–424.

85. Rowan, M.; Mamel, K.; Tuovinen, O. H. Clogging Incidence of Drip Irrigation Emitters Distributing Effluents of Differing Levels of Treatment. In On-Site Waste Water Treatment X, Conference Proceedings, 2004; p 084–091.

86. Dehghanisanij, H.; Yamamoto, T.; Ould Ahmad, B.; Fujiyana, H.; Miyamoto, K. The effect of chlorine on emitter clogging induced by algae and protozoa and the performance of drip irrigation. *Trans. Am. Soc. Agric. Eng.* 2005, 48, 519–527.

87. Barges, P. J.; Arbat, G.; Barragan, J.; Ramirej de Cartogene, F. Hydraulic performance of drip irrigation subunits using WWTP effluents. *Agric. Water Manage.* 2005, 77, 249–262.

88. Barges, P. J.; Barragan, F.; Ramirez, D. C. Filtration of effluents for micro irrigation systems. *Trans. ASAE*, 2005, 48(3), 969–978.

89. Yuan, Z.; Waller, P. M.; Choi, C. Y. Effect of organic acids on salt precipitation in drip emitters and soil. *Trans. Am. Soc. Agric. Eng.* 1998, 41(6), 1689–1696.

90. Elisa, M.; Suarez-Rey, K. Effects of chemicals on root intrusion into subsurface drip emitters. *J. Irrig. Drain.* 2006, 55, 501–509.

91. El-Berry, A. M.; Bakeer, G. A.; Al-Weshali, A. M. *The Effect of Water Quality and Aperture Size on Clogging of Emitters.* 2007. http://www.sciencedirect.com.

92. Chang, C. Chlorination for disinfection and prevention of clogging of drip lines and emitters. *Encycl. Water Sci.* 2008, 26.

93. Duran-Ros, M.; Puig-Bargues, J.; Arbat, G.; Barragan, J.; Ramirez de Cartagene, F. Effect of filter, emitter and location on clogging when using effluents. *Agric. Water Manage.* 2009, 96, 67–79.

94. Liu, H.; Huang, G. Laboratory experiment on drip emitter clogging with fresh water and treated with sewage effluent. *J. Agric. Water Manage.* 2009, 96, 745–756.

95. Li, J.; Chen, L.; Li, Y. Comparison of clogging in drip emitters during application of sewage effluent and groundwater. *Trans. Am. Soc. Agric. Biol. Eng.* 2009, 52(4), 1203–1211.

96. Barges, P. J.; Arbat, G.; Elbana, M.; Duran-Ros, M.; Barragan, J.; Ramirej de Cartogene, F.; Lamm, F. R. Effect of flushing frequency on emitter clogging in micro irrigation system with effluents. *Agric. Water Manage.* 2010, 97(6), 883–891.

97. Yan, Z. B.; Mike, R.; Likun, G. U.; Ren, S.; Peiling, Y. Biofilm structure and its influence on clogging in drip irrigation emitters distributing reclaimed wastewater. *J. Environ. Sci.* 2010, 21, 834–841.

98. Felix, B. R. Performance of Drip Irrigation Systems Under Field Conditions; ARC-Institute for Agricultural Engineering: Silverton, South Africa, 2010.

99. Raposo, M.; Osvaldo, N. *Adsorption Mechanisms in Layer-By-Layer Films*; Universidade de Sao Paulo: Sao Carlos, SP, Brasil, 1998, CP 369-13560-970.

100. Maarof, H. I.; Bassim, H. H.; Abdul, L. A. Adsorption Isotherm of Q-Cresol from Aqueous Solution by Granular Activated Carbon. In Proceedings of International Conference on Chemical and Bioprocess Engineering; Universiti Malaysia Sabah: Kola Kinabalu, 2003; p 21 – 29.

101. Alves, O.; Lavorenti, A. Sulfate adsorption and its relationships with properties of representative soils of the Sao Paulo State, Brazil. *Geoderma*, 2003, 118, 89–99.

102. Hussain., A.; Anwar-Ul- Haq, G.; Muhammad, N. Application of the Langmuir and Freundlich equations for P adsorption in saline-sodic soils. *Int. J. Agric. Biol.* 2003, 5(3), 91–99.

103. Agrawal, K.; Sahu, K.; Pandey, B. D. A comparative adsorption study of copper on various industrial solid wastes. *AICHE J.* 2004, 50(10), 34–41.

104. Siswati, G.; Uday, G. Studies on adsorption of Cr (VI) onto synthetic hydrousstannic oxide. *Water SA*, 2005, 31(4), 21–25.

105. Chattha, K.; Yousaf, M.; Javeed, S. Phosphorus adsorption as described by freundlich adsorption isotherms under rainfed conditions of Pakistan. *Pak. J Agric. Sci.* 2007, 44(4), 27.

106. David, L. M.; Jeffrey J. G. Solution and adsorbed state structural ensembles predicted for the Statherin hydroxyapatite system. *J. Biophys.* 2009, 96(8), 3082–3091.

107. Laizhou, S.; Wang, J.; Zheng, Q.; Zhang, Z. Characterization of Cu (II) ion adsorption behavior of the polyacrylic acid-polyvinylidene fluoride blended polymer. 2008.

108. Ashraf, M. S.; Ranjha, A. M.; Yaseen, M.; Ahmad, N.; Hannan, A. Zinc adsorption behaviour of different textured calcareous soils using Freundlich and Langmuir models. *Pak. J. Agric. Sci.* 2008, 45(1), 83–89.

109. Shafiq, R.; Yaseen, A. M.; Mehdi, M. S. M.; Hannan, A. Comparison of Freundlich and Langmuir adsorption equations for boron adsorption on calcareous soils. *J. Agric. Res.* 2008, 46(2), 9.

110. Dai, J. L.; Zhang, M.; Hu, Q. H.; Huang, Y. Z.; Wang, R. Q.; Zhu, Y. G. Adsorption and desorption of iodine by various Chinese soils, II. Iodide and iodate. *Geoderma*, 2008, 153, 130–135.

111. Kozak, P. P.; Vladimir, M.; Zdenik, K. Adsorption of zinc and copper ions on natural and ethylenediamine modified montmorillonite. *Ceram. Silik.* 2010, 54(1), 78–84.

112. Claesson, P. M.; Makuska, R. I.; Varga, R.; Meszaros, S.; Titmuss, P.; Linse, K. Bottle brush polymers and surface adorption and their interactions. *Adv. Colloid Interface Sci.* 2010, 155, 50–57.

113. Zhang, J.; Zhao, W.; Tang, Y.; Wei, Z.; Lu, B. Numerical investigation of the clogging mechanism in labyrinth channel of the emitter. *Int. J. Numer. Methods Eng.* 2006, 70, 1598–1612.

114. Boparai, H. K.; Meera, J.; Denis, M.; Carroll, O. Kinetics and thermodynamics of cadmium ion removal by adsorption onto nano zerovalent iron particles. *J. Hazard. Mater.* 2011, 186, 458–465.

115. Osunbitan, J. A.; Adekalu, K. O.; Aina, P. O. Adsorption behaviour of manganese from Dithane M– 45 to two soil types in Southwestern Nigeria. *Can. J. Environ. Constr. Civil Eng.* 2011, 2(5), 77–81.

116. Chavan, V. K.; Balakrishnan, P.; Deshmukh, S. K.; Ingle, V. K. Evaluation of adsorption mechanism in clogging of materials used in drip irrigation system. *J. Agric. Eng.* 2013, 50(3).

117. Chavan V. K.; Balakrishnan, P.; Deshmukh, S. K.; Nagdeve, M. B. *Chapter 6: Mechanics of Clogging in Micro Irrigation System.* In *Sustainable Practices in Surface and Subsurface Micro Irrigation*; Goyal, M. R. Ed.; Apple Academic Press Inc.: Oakville, ON, Canada, 2014; Vol. 2; p 169–182.

118. Capra, A.; Scicolone, B. Recycling of poor quality urban wastewater by drip irrigation systems. *J. Clean. Prod.* 2007, (16).

119. Government of India. Report of the Working Group on Water Resources For the XI the Five Year Plan (2007–2012); Ministry of Water Resources. 2006; p 54.

120. Hills, D. J.; Navar, F.M.; Waller, P.M. Effects of chemical clogging on drip-tape irrigation uniformity. *Trans. Am. Soc. Agric. Eng.* 1989, 32(4), 1202–1206.

121. Ahmad Aali, K.; Liaghat, A.; Dehghanisanij, H. The Effect of acidification and magnetic on emitter clogging under saline water application. *J. Agric. Sci.* 2009, 1(1), 21–24.

122. Nakayama, F. S. Water treatments in trickle irrigation systems. *J. Irrig. Drain. Div. Am. Soc. Civil Eng.* 1978, 23–24.

123. Narayanamoorthy, A. *Efficiency of Irrigation: A Case of Drip Irrigation. Occasional Paper: 45*; Department of Economic Analysis and Research, National Bank for Agriculture and Rural Development: Mumbai, India, 2005.

124. Zhang, S. H.; Jing, H.; Fuxing, G.; Yuh-Shan, H. *Adsorption Characteristics Studies of Phosphorous onto Literate.* In *Desalinization and Water Treatment* 1944 994/1944-3986; Desalination Publications, 2010; p 98–105.

PART II

MANAGEMENT OF EMITTER CLOGGING

CHAPTER 3

INCIDENCE OF CLOGGING UNDER SURFACE AND SUBSURFACE DRIP IRRIGATION WITH GROUNDWATER

VINOD KUMAR TRIPATHI

CONTENTS

3.1   INTRODUCTION

Drip irrigation provides water with uniform rate, precisely controls the amount of water, increases crop yield, reduces evapotranspiration and deep percolation, and decreases dangers of soil degradation and salinity.[1–3] Well-aerated conditions (appropriate proportion of soil and water) can also be maintained. In addition, drip irrigation can supply water at low discharge rates and high frequencies over an extended period and minimize salinity levels in the soil water by leaching the salts.[4] Because of point-source characteristics of drip irrigation, salts will be pushed toward the fringe of the wetting area along with water and can form a desalinization zone at its center, in close proximity to the emitter.[5,6] Many studies have shown that drip irrigation on saline soil is an important method for improving saline land.[7–9]

Drip irrigation system applies precise amount of water to the crop at the right time and ensures its uniform distribution in the field. Although it is the most efficient method, the suspended salts and solids in groundwater (GW) can lead to a high risk of system failure due to clogging of the drippers and difficulties in filtering. In drip irrigation system, quality of water, emitter characteristics, and filter efficacy play key roles to minimize clogging. However, other factors being the same, the most important feature for success is good filtration system.[10,11]

In India, gravel media filters, screen filters, and disk filters are commonly used to remove sediments from water for drip irrigation. According to Capra and Scicolone,[12] screen filters are not suitable for use with wastewater, the exception being diluted and settled wastewater. They also observed almost similar performance by disk and gravel media filters with treated municipal wastewater. Besides, many researchers have conducted studies with freshwater and wastewater using drip irrigation mostly by surface placement of laterals and also under laboratory conditions.[12–15] Subsurface drip irrigation (SDI) has not been evaluated to use GW. Thus, there is a need to develop a methodology for utilizing GW through drip irrigation on sustainable basis.

Therefore, the present study was conducted under actual field situations using surface and subsurface drip with two types of filters so that some guidelines can be suggested on use of drip irrigation with GW.

3.2   MATERIALS AND METHODS

3.2.1   WATER RESOURCE

The field experiments were conducted at Precision Farming Development Centre of Water Technology Centre, IARI, Pusa, New Delhi. Randomized block statistical design was used. The GW was collected from the tube-well installed at Indian Agricultural Research Institute (IARI), New Delhi, India. Water samples were analyzed for pH, electrical conductivity (EC), total solids (dissolved and undissolved), turbidity, calcium, magnesium, carbonate, and bicarbonate according to the standard methods.[16]

3.2.2 EXPERIMENTAL SETUP

Drip irrigation system was installed as shown in Fig. 3.1. In-line lateral (J-Turbo Line) with 40 cm emitter spacing was laid on the ground for surface drip and was buried at a depth of 15 and 30 cm from ground surface for subsurface drip. Filtration system included sand media filter (F1: flow rate 30 $m^3 \cdot h^{-1}$, 50 mm size, silica sand 1.0–2.0 mm, thickness 80 cm) with back flush mechanism and disk filter (F2: flow rate 30 $m^3 \cdot h^{-1}$, 20 mm size, 130 μm, disk surface 1.198 cm^2, screen surface 815 cm^2 AZUD helix system, model 2NR). Water was allowed to pass through the combination of filters F1 and F2. Main line (50 mm diameter, PVC pipe) was connected to sub-mains (35 mm diameter, PVC pipe) for each of the plots through a gate valve.

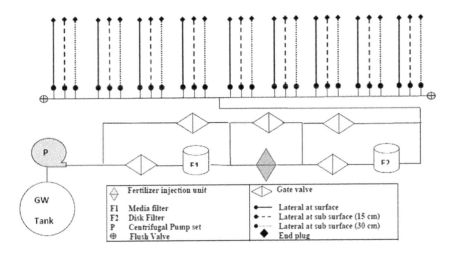

FIGURE 3.1 Layout of the field experiment.

3.2.3 OPERATIONAL PROCEDURE

The GW collected from the tube-well was stored in a tank (Fig. 3.1). Water from the tank was fed to the filtration system and then allowed to pass through emitters. The pump was turned on, and emitters were allowed to drip for approximately 2 min to allow air to escape. The water collection period was set at 5 min. Quantity of flow of water from drip emitter was collected in containers at 98.06 kPa pressure and was repeated three times.

The flow rate was estimated by dividing the total volume collected by the time of collection. The measurement was taken from randomly located sampling emitters for the determination of parameters of performance evaluation. Discharge from SDI laterals were measured by excavating the soil around the buried drip laterals so that an emitter is visible with sufficient space below it for the placement of the container to collect water from the emitter as suggested by Camp et al.[17] and Magwenzi.[18] Performance of the system was evaluated at normal operating pressure to discharge sufficient water for infiltration and to avoid ponding near the emitters. According to the recommendation of the manufacturers, an operating pressure of 98.06 kPa was considered appropriate. To achieve accurate pressure, emitter-level measurement was done at the lateral with digital pressure gauge having the least count of 0.01 kPa.

3.2.4 PARAMETERS OF PERFORMANCE EVALUATION

3.2.4.1   HEAD–DISCHARGE RELATIONSHIP OF EMITTERS

A numerical description of pressure flow characteristics for a given emitter device is based on the flow rate versus pressure curve fitted to an equation of the following form:

$$CV_q = \frac{SD}{q}100 \tag{1}$$

where is the emitter flow rate ($m^3 \cdot s^{-1}$); is the emitter coefficient that accounts for real discharge effects and makes the units correct ($L \cdot s^{-1}$); is the pressure head in the lateral at the location of emitters (m), and is the exponent characteristic of the emitter (dimensionless).

The exponent indicates the flow regime and emitter type and typically ranges between 0.0 and 1.0. This exponent is a measure of flow rate sensitivity to pressure change. A higher value for indicates higher sensitivity. The emitter exponent and constant value were derived using polynomial regression.

3.2.4.2   COEFFICIENT OF VARIATION

The coefficient of variation (CV) of the emitter discharge in a lateral was calculated[19,20] using the following equation:

$$CV_q = \frac{SD}{q}100$$
(2)

where SD is the standard deviation of emitter discharge (L·h⁻¹) and   is the mean discharge in the same lateral (L·h⁻¹). Minimum value of CV was observed at 98.06 kPa pressure. As such, it was selected as the operating pressure for further study of clogging.

3.2.4.3. EMITTER FLOW RATE (% OF INITIAL)

The emitter flow rate (% of initial,) was calculated as follows:

$$R = \frac{q}{q_{ini}}100$$
(3)

where   is the mean emitter discharges of each lateral (L·h⁻¹), and $_{ini}$ is the corresponding mean discharge of new emitters at the same operating pressure of 98.06 kPa (L·h⁻¹).

3.2.4.4   UNIFORMITY COEFFICIENT

Uniformity coefficient by Christiansen[21] was calculated as follows:

$$UC = 100\left[1 - \frac{\frac{1}{n}\sum_{i=1}^{n}|q_i - q|}{q}\right]$$
(4)

where $_i$ is the measured discharge of emitter i (L·h⁻¹), is the mean discharge at drip lateral (L·h⁻¹), and is the total number of emitters evaluated.

3.2.4.5. VARIATION IN FLOW RATE (Q_{VAR})

Emitter flow rate variation $_{var}$[20] was calculated using the following equation:

$$q_{var} = \frac{q_{max} - q_{min}}{q_{max}}$$
(5)

where $_{max}$ is the maximum flow rate (L·h^{-1}) and $_{min}$ is the minimum flow rate (L·h^{-1}).

3.2.5 STATISTICAL ANALYSIS

Statistical analysis was carried out using the GLM procedure of the SAS statistical package (SAS Institute, Cary, NC, USA). The model used for analysis of variance (ANOVA) included water from different filters and placement of lateral as fixed effect and interaction between filtered water and depth of emitter. The ANOVA was performed at probabilities of 0.05 or less level of significance to determine whether significant differences existed among treatment means.

3.3 RESULTS AND DISCUSSIONS

3.3.1 CHARACTERIZATION OF WATER

The physical and chemical characteristics of GW are presented in Table 3.1. It was observed that the EC value for the GW was from 1.89 to 2.58 dS·m^{-1} with an average of 2.16 dS·m^{-1}. Higher EC values indicate that salt content in the GW contributes more to chemically induced emitter clogging. Variation in pH values was 6.95–8.57 with a mean value of 7.40 indicating slight basic nature of GW. The pH may not have direct impact on clogging, but it can accelerate the chemical reactions or biological growth involved in clogging.[22,23] Variation in the values of total solids was in the range of 800–1533 mg·L^{-1} with a mean value of 967 mg·L^{-1}. Highest value of total solids (1533 mg·L^{-1}) was observed in the month of May.

TABLE 3.1 Physicochemical and Biological Properties of Water Used for Irrigation

Properties	Units	Groundwater
		Mean ± SD
Ca	mg·L^{-1}	44.58 ± 8.27
CO$_3$	mg·L^{-1}	58.0 ± 8.23
EC	dS·m^{-1}	2.17 ± 0.25
HCO$_3$	mg·L^{-1}	364.33 ± 70.7
Mg	mg·L^{-1}	35.28 ± 5.81
pH	–	7.4 ± 0.43
Total solids	mg·L^{-1}	967.4 ± 212.6
Turbidity	NTU	1.50 ± 0.52

Turbidity had negligible values with a maximum of only 2 NTU. Variation in calcium content of GW was in the range of 36–66 mg·L^{-1} with an average of 45 mg·L^{-1}. Variation in magnesium content was in the range of 23–42 mg·L^{-1} with a mean value of 36 mg·L^{-1}. Carbonate content of GW was in the range of 48–78 mg·L^{-1} with a mean value of 58 mg·L^{-1}. Variation in bicarbonate content of GW was in the range of 264–496 mg·L^{-1} with a mean of 364 mg·L^{-1}. The presence of carbonate was lower in comparison with bicarbonate. This also gives an indication of the presence of magnesium carbonate in GW. On the basis of these quality parameters, it can be concluded that clogging problem could be encountered with the GW through drip irrigation system.

3.3.2 EMITTER HYDRAULICS

Coefficients for equation (eq 1) were decreased with the time of operation of emitters in all filtration systems as a result of partial clogging (Table 3.2). Theoretically, the exponent for the emitter was 0.5, which comes under the category of completely turbulent hydraulic regime.[24] In normal operating pressure range, exponent was less than 0.5 for all depths of placement of drip laterals. But the difference was higher under subsurface placement of drip lateral at 30 cm depth. Similar trend was observed for discharge coefficient (). The coefficient of regression (^{2}) was 0.99 for all situations, and this indicated that the equation described the flow–pressure relationship precisely.

TABLE 3.2 Relationships for Emitter under Combination Filtration System

Filter	Placement of Lateral	Stage	Coefficient, C	Exponent, x	R^2
Combination of F1 and F2	Surface	Beginning	3.548	0.494	0.99
		Middle	3.446	0.494	0.99
		End	3.350	0.493	0.99
	Subsurface (15 cm)	Beginning	3.548	0.494	0.99
		Middle	3.431	0.494	0.99
		End	3.350	0.493	0.99
	Subsurface (30 cm)	Beginning	3.548	0.494	0.99
		Middle	3.415	0.493	0.99
		End	3.298	0.492	0.99

3.3.3   COEFFICIENT OF VARIATION OF EMITTER DISCHARGE (CV_Q)

The CV of the discharge for the combination of both filters is presented in Fig. 3.2. In the beginning, maximum CV of 1.25% was observed under surface and subsurface placement of drip laterals. As shown in Fig. 3.2, after 1 year (beginning) in the entire depth of placement, CV was less than 5%. Hence, the performance can be rated as excellent.[25] But after 2 years of experimentation (end of experiment), maximum variation of 9.5% was observed under subsurface placement of drip laterals at 15 cm depth. Maximum deviation of 5.06% was observed under subsurface placement of drip laterals at 30 cm depth. The results indicate that 1-year operation of the emitters did not cause much variation, but continuous 2 years of operation caused significant variation in emitter discharge. This is also supported by the computation of the standard error, which was lower in the first year but was significantly higher in the second year under entire depth situation. Coefficient of variation in subsurface placement condition was always poor than surface placement.

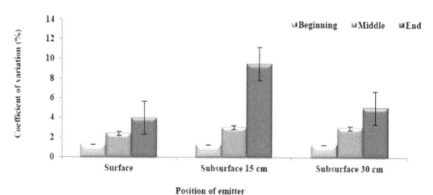

FIGURE 3.2   Coefficient of variation in emitter discharge under different filtration systems at 98.06 kPa pressure.

3.3.4   EMITTER FLOW RATE

Maximum reduction in flow rate was observed under subsurface placement of drip laterals at 30 cm depth and minimum was observed under surface placement of drip laterals. Reduction in discharge was within these two values (Fig. 3.3). The statistical analysis revealed that after 2 years of experiment, there was a significant effect of filter, emitter placement, and their interaction on the discharge of drip emitters (Table 3.3). In the beginning of the experiment, there was no significant effect of emitter placement and their interaction with filtered water because emitters were new and there was no clogging. After continuous use, clogging takes place, and the effect of different filtration systems starts showing up in the discharge of emitters. At the end of 1 year, the effect of filtration system was significant, but the effect of emitter placement was not significant. Both were significant after 2 years of use. These results indicated that clogging is a dynamic phenomenon over time.[26]

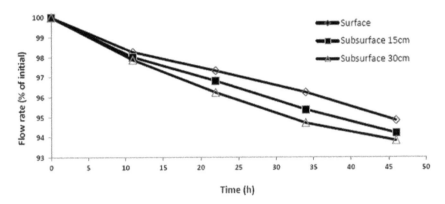

FIGURE 3.3   Emitter flow rate (% of initial flow rate) under different filtration systems for the GW at 98.06 kPa pressure.

TABLE 3.3   Significance Level (-Value) of the Statistical Model and of Each Factor and Interaction for Emitter Flow Rate

Parameter	Time		
	Beginning	**1 Year**	**2 Years**
Model	$(^2 = 0.97)$*	$(^2 = 0.87)$**	$(^2 = 0.93)$*
Filter (F)	NS	**	*
Emitter placement (EP)	NS	NS	***
F × EP	NS	**	*

Note: NS, not significant, > 0.05; * < 0.001; ** < 0.01; *** < 0.05.

3.3.5 UNIFORMITY OF WATER APPLICATION

Variations in uniformity coefficient and flow rate are presented in Table 3.4. Least variation in flow rate with maximum uniformity was observed at the beginning of the experiment. Variation in flow rate was increased with the operation of drip system, and maximum variation with minimum uniformity coefficient was reached at the end of 2 years of experimentation. Performance of filter combination could be rated as good.[27] After 2 years, minimum uniformity coefficient was 92.54 under subsurface placement of drip laterals at 30 cm depth. As per general criteria for $_{var}$, values of 0.10 or less are desirable, 0.1–0.2 are acceptable, and greater than 0.2 are unacceptable. Two out of three depths gave variation in flow rate under acceptable limit.

TABLE 3.4 Uniformity Coefficient and Variation in Flow Rate ($_{var}$) Resulting from the Performance Evaluation of Drip Irrigation System

Filter	Depth of Placement of Lateral	Uniformity Coefficient			Variation in Flow Rate (q_{var})		
		Beginning	1 Year	2 Years	Beginning	1 Year	2 Years
Combination of filters	Surface	99.07	97.29	95.55	0.048	0.102	0.131
	Subsurface 15 cm	99.02	96.02	93.50	0.048	0.100	0.160
	Subsurface 30 cm	99.07	94.71	92.54	0.048	0.126	0.211

3.4 CONCLUSIONS

The hydraulic performance of the drip emitters revealed that for continuous use of the GW, filtration with a combination of gravel and disk filter is the most appropriate strategy against emitter clogging. It resulted in a better emitter discharge exponent, a reasonably good coefficient of variation, and uniformity coefficient. Performance of emitters was lower under subsurface placement of drip laterals in comparison with subsurface placement of drip laterals.

3.5 SUMMARY

Drip irrigation is the most efficient irrigation method because it applies water precisely and uniformly at high frequencies, maintaining high soil matric potential in the root zone for crop growth. Scarcity of freshwater is putting a lot of pressure on irrigation engineers for its judicious use in agriculture. Utilization of GW for irrigation through drip irrigation system is the best choice to reduce the demand of freshwater under irrigation sector. Since clogging is the major problem associated with GW utilization through drip irrigation system.

Physical and chemical characteristics of the GW were determined, and the effect of water quality on the performance of emitters was evaluated. Although higher EC, pH, Mg, and carbonate were observed, lower turbidity, total solids, and HCO_3 were also observed. Surface placement of drip laterals in comparison with subsurface placement of drip laterals gave emitter discharge exponent close to 0.5 with an [2] value of 0.99. Emitter flow rate was decreased with the increase in time of operation of the system. Coefficient of variation less than 4% showed excellent performance in surface-placed drip lateral after 2 years of operation. After 2 years, coefficient of variation (CV) of 9.5% was observed under subsurface placement of laterals at 15 cm depth, but 5.1% was observed under subsurface placement of laterals at 30 cm depth.

KEYWORDS

- characterization of water
- coefficient of variation
- crop growth
- dripper
- emitter
- emitter discharge
- emitter discharge exponent
- emitter flow rate
- emitter hydraulics
- emitter placement
- filter
- groundwater
- head–discharge relationship
- hydraulic coefficient
- lateral placement
- micro irrigation
- subsurface drip irrigation, SDI
- surface drip irrigation
- uniformity coefficient

REFERENCES

1. Ayars, J. E.; Phene, C. J.; Hutmacher, R. B.; Davis, K. R.; Schoneman, R. A.; Vail, S. S.; Mead, R. M. Subsurface drip irrigation of row crops: a review of 15 years of research at the Water Management Research Laboratory. 1999, 42, 1–27.
2. Batchelor, C. H.; Lovell, C. J.; Murata, M. Simple micro irrigation techniques for improving irrigation efficiency on vegetable gardens. . 1996, 32, 37–48.
3. Karlberg, L.; Frits, W. T. P. V. Exploring potentials and constraints of low-cost drip irrigation with saline water in sub-Saharan Africa. , 2004, 29, 1035–1042.
4. Keller, J.; Bliesner, R. D. Ed., ; Van Nostrand Reinhold: New York, 1990; p 22.
5. Goldberg, D.; Gornat, B.; Rimon, D. Ed.; Drip Irrigation Scientific Publications: Israel, 1976.
6. Kang, Y. H. Micro irrigation for the development of sustainable agriculture. , 1998, 14 (Suppl.), 251–255.
7. Chen, M.; Kang, Y. H.; Wan, S. Q.; Liu, S. P. Drip irrigation with saline water for oleic sunflower (L.). . 2009, 96, 1766–1772.
8. Kang, Y. H.; Chen, M.; Wan, S. Q. Effects of drip irrigation with saline water on waxy maize (Zea mays L. var. ceratina Kulesh) in North China Plain. 2010, 97, 1303–1309.
9. Wan, S. Q.; Kang, Y. H.; Wang, D. Effect of drip irrigation with saline water on tomato (Lycopersicon esculentum Mill) yield and water use in semi-humid area. . 2007, 90, 63–74.
10. McDonald, D. R.; Lau, L. S.; Wu, I. P.; Gee, H. K.; Young, S. C. H. Improved Emitter and Network System Design for Reuse of Wastewater in Drip Irrigation. In Technical Report 163; Water Resources Research Centre, University of Hawaii at Manoa: Honolulu, 1984.
11. Oron, G.; Shelef, G.; Turzynski, B. Trickle irrigation using treated wastewaters. . 1979, 105(IR2), 175–186.
12. Capra, A.; Scicolone, B. Emitter and filter test for wastewater reuse by drip irrigation. . 2004, 68(2), 135–149.
13. Cararo, D. C.; Botrel, T. A.; Hills, D. J.; Leverenz, H. L. Analysis of clogging in drip emitters during wastewater irrigation. , 2006, 22(2), 251–257.
14. Liu, H.; Huang, G. Laboratory experiment on drip emitter clogging with fresh water and treated sewage effluent. . 2009, 96, 745–756.
15. Rowan, M.; Manci, K.; Tuovinen, O. H. Clogging Incidence of Drip Irrigation Emitters Distributing Effluents of Different Levels of Treatments. In Conference Proceeding on On-Site Wastewater Treatment, Sacramento (California), USA, March 21-24, 2004, p 84–91, 2004.
16. APHA. ; American Public Health Association: Washington, DC, 2005.
17. Camp, C. R.; Sadler, E. J.; Buscher, W. J. A comparison of uniformity measures for drip irrigation systems. , 1997, 40(4), 1013–1020.
18. Magwenzi, O. Efficiency of Subsurface Drip Irrigation in Commercial Sugarcane Field in Swaziland. 2001, p 1–4. http://www.sasa.org.za/sasex/about/agronomy/aapdfs/ magwenzi. pdf.
19. Bralts, F. V.; Kesner, D. C. Drip irrigation field uniformity estimation. , 1983, 26, 1369–1374.
20. Wu, I. P.; Howell, T. A.; Hiler, E. A. Hydraulic design of drip irrigation systems. Hawaii Agric Exp Stn. Tech. Bull. 105, Honolulu, 1979.
21. Christiansen, J. E. Hydraulics of sprinkler systems for irrigation. , 1942, 107, 221–239.
22. Dehghanisanij, H.; Yamamoto, T.; Rasiah, V.; Utsunomiya, J.; Inoue, M. Impact of biological clogging agents on filter and emitter discharge characteristics of micro irrigation system. . 2004, 53, 363–373.
23. Nakayama, F.S.; Bucks, D.A. Water quality for drip/trickle irrigation: a review. . 1991, 12, 187–192.
24. Cuenca, R. H. ; Prentice-Hall: Englewood Cliffs, New Jersey, 1989; p 317–350.

25. ASABE. .; American Society of Agricultural and Biological Engineers: St. Joseph, Michigan, 2003.
26. Ravina, I.; Paz, E.; Sofer, Z.; Marcu, A.; Shisha, A.; Sagi, G. Control of emitter clogging in drip irrigation with reclaimed wastewater. . 1992, 13(3), 129–139.
27. Puig-Bargues, J.; Arbat, G.; Barragan, J.; Ramirez de Cartagena, F. Hydraulic performance of drip irrigation subunits using WWTP effluents. . 2005, 77(1–3), 249–262.

CHAPTER 4

CHALLENGES IN CLOGGING UNDER SUBSURFACE DRIP IRRIGATION: SOUTH AFRICA

FELIX B. REINDERS

CONTENTS

4.1   INTRODUCTION

South Africa is a dry country with a rainfall below world average, which is distributed unequally throughout the country. This rainfall is also highly irregular in occurrence, and the demand for water has created pressure for the optimal use of all water. Therefore, many farmers invest in drip irrigation as an improved or most efficient irrigation method for water conservation. However, there is proof from literature and from this research that this system can also be inefficient as a result of water quality, mismanagement, clogging, and maintenance problems.

The South African National Water Act (Act 36 of 1998) makes provision for water to be protected, used, developed, conserved, managed, and controlled in a sustainable and equitable manner to the benefit of all people in South Africa.[1,2] Currently, subsurface drip systems (SDI) account for ±7,500 hectares of the total of 140,000 hectares under drip irrigation in South Africa out of a total of 1.3 million-ha.

To assist the users to utilize the systems effectively, this research was carried out to determine the blockage potential of different types of emitters under field and laboratory conditions.

4.2   METHODOLOGY

Research was carried out by the Agricultural Research Council – Institute for Agricultural Engineering (ARC-IAE) on three drip irrigation equipments (by different suppliers of drip irrigation systems) to determine the blockage potential due to root intrusion. Research was carried out in an experimental field on five different types of emitters, and retrieved dripper lines from the experimental field were tested in the laboratory. Evaluations were also carried out in the field under farming conditions.

First, an extensive literature study on all facets that can influence different types of emitters under field conditions was undertaken. Aspects that were addressed in this study include water quality, water treatment methods, inherent factors that affect emitter performance, filtering, chemical treatment of the soil surrounding the dripper lines, system maintenance, system installation, and design.

4.2.1   SELECTION OF DRIPPERS

Agriplas, Netafim, and T-Tape companies (Table 4.1) were selected to determine the blockage potential of different types of emitters using five models.

TABLE 4.1   Five Types of Emitters under Study

Emitter Name (Manufacturer)	Q, lph	Flow Path[a]	Lateral Type[b]	Aperture Type	Zone[c] (Code)
DIS PC Lite (Agriplas)	2.2	Turbulent pressure compensated	Tube, thick wall	2 × 2 mm Ø circular aperture	8 (A400)
Dripin Light (Agriplas)	1.7	Turbulent	Tube, thick wall	2 × 3 mm Ø circular aperture	5 (A500)
Ram 17L (Netafim)	1.6	Turbulent pressure compensated	Tape, thin wall	3 mm Ø circular aperture	6 (A200)
Super Typhoon (Netafim)	1.7	Turbulent	Tape, thin wall	1.8 mm Ø circular aperture	4 (A300)
T-Tape (T-Tape)	1.0	Turbulent	Tape, thin wall	32 mm longitudinal slit	7 (A100)

Note: Q is the Delivery rate, nominal.

[a]Flow path: *turbulent flow path*, this is when obstacles are placed in the flow path of the water in an emitter, to reduce kinetic energy in the water to change it from laminar flow to turbulent flow; *pressure compensated*, this is when a diaphragm is introduced in an emitter, depending on the pressure in the dripper line, to regulate the area of the flow path to maintain a constant flow rate.

[b]Lateral type: *tape, thin wall*, emitter lines with a wall thickness of 0.5 mm or less; *tube, thick wall*, emitter lines with a wall thickness more than 0.5 mm.

[c]Zone refers to a group of experimental blocks of a specific flow path.

4.2.2   LABORATORY TESTS ON DRIPPERS

The new drip lines with emitters were tested under controlled conditions in the Hydro Laboratory of ARC-IAE for average discharge ($\bar{q}$) and for the manufacturing coefficient of discharge variation (CV_q). These values were used as a reference base in the evaluation of the experimental site and infield performance of the particular emitter types. Both $\bar{q}$ and CV_q were determined for a total sample of 100 emitters, as well as for four groups of 25 emitters in accordance with the International Standards Organisation[3] and expressed as in eqs 1–3:

$$\bar{q} = \frac{1}{n}\sum_{i=1}^{n} q_i \tag{1}$$

$$S_q = \left[\frac{1}{n-1} \sum_{i=1}^{n} (q_i - \bar{q})^2 \right]^{\frac{1}{2}} \qquad (2)$$

$$CV_q = \frac{S_q}{\bar{q}} \times 100 \qquad (3)$$

where q_i is the emitter discharge rate (L/h); n is the number of emitters of the sample; $\bar{q}$ is the mean of all the measured discharge rates (L/h); S_q is the standard deviation of the discharge rate of the emitter; and CV_q is the coefficient of variation of the discharge rate of the emitters (%).

The coefficient of manufacturing variation (CV_q) is used as a measure of the anticipated variation in discharge for a sample of new emitters. The CV_q is a very useful parameter with rather consistent physical significance, because the discharge rate for emitters at a given pressure is essentially normally distributed. Criteria for CV_q are given in Table 4.2.

TABLE 4.2   Criteria for CV_q (%) of "Point-Source" Drippers

Classification	ASAE EP 405.1 (1997)	Classification	ARC-IAE	ISO
Excellent	<5	Excellent	0.1–2.5	
Average	5–7	Good	2.6–5.0	0.1–5.0
Marginal	7–11	Fair	5.1–7.5	
Poor	11–15	Marginal	7.6–10	5.1–10
Unacceptable	>15	Poor	>10	>10

Note: The physical significance of CV_q is derived from the classic bell-shaped normal distribution curve as shown in Fig. 4.1.

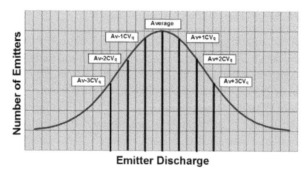

FIGURE 4.1   Normal distribution curve.

In a normal distribution, the following conditions are satisfied:

- Essentially all the observed discharge rates fall within $(1 \pm 3CV_q)$ of the average discharge rate.
- Approximately 95% of the discharge rates fall within $(1 \pm 2CV_q)$ of the average discharge rate.
- Approximately 68% of the discharge rates fall within $(1 \pm 1CV_q)$ of the average discharge rate.
- The average of the lowest quarter of the discharge rates is approximately = $(1 - 1.27CV_q)$ of the average discharge rate.

4.2.3   TESTS ON EMITTERS RETRIEVED FROM THE EXPERIMENTAL SITE

To evaluate the impact of root intrusion, a statistical-based design and layout of 40 blocks with five dripper types was installed at a site at ARC-IAE. All blocks were managed according to acceptable norms, and 20 of them were managed by applying a root growth inhibitor (Treflan) with the purpose to evaluate the prevention of root intrusion. Laterals were retrieved from the experimental site after 19, 30, and 42 months for testing in the laboratory.

The test results on the retrieved emitters are discussed in terms of functional and clogged emitters. "Functional emitters" are those emitters that delivered water, independent of flow rate, with a delivery rate more than zero liters per hour. The variance in discharge rates (delivery rates) of the functional emitters is discussed in the relevant paragraphs under discharge categories. The functional emitters were divided into three discharge categories, namely, the *reduced discharge*, the *average discharge*, and the *increased discharge*. The categories are based on the CV_q values of the tests of the specific dripper laterals.

Clogged emitters are discussed in terms of root clogged and other clogged emitters. "Clogged emitters" indicate either root clogged or other clogged with the following definition:

- *Root clogged emitters* indicate emitters that were clogged by roots, as observed in the orifices or in the labyrinths or dripper mechanisms.
- *Other clogged emitters* indicate those emitters clogged by soil, organic matter, for example, algae or any foreign matter, excluding roots.

4.2.4   EMITTER DISCHARGE CATEGORIES

In a further analysis to understand the impact of clogging better, the data of different zones were divided into different discharge categories. The tested emitter discharge values of retrieved emitters were categorized according to the CV_q and discharge values of the similar new dripper lines.

The discharges of functional emitters were tabled into three different categories. A graphical illustration of the categories can be seen in Fig. 4.2. The categories were defined as follows:

- Reduced discharge: The "reduced discharge category" is where the discharge rates of the emitters fall below the discharge rate of Av *minus* $2 \times (CV_q)$.
- Average discharge: The average discharge is written as **Av** and the "coefficient of variation" as CV_q (Fig. 4.1). The "average discharge category" is where the discharge values fall between Av *minus* $2 \times (CV_q)$ and Av *plus* 2 (CV_q).
- Increased discharge: The "increased discharge category" is where the discharge values are more than Av *plus* $2 \times (CV_q)$.

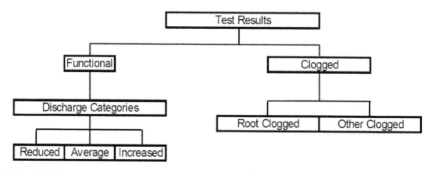

FIGURE 4.2   Graphical structure of test results for discussion.

4.2.5 FIELD EVALUATION OF DRIP SYSTEMS

With regard to the field evaluation, two areas in Southern Africa were identified: Mpumalanga Lowveld and Swaziland. In these areas, a total of five systems were identified. The performance of these systems was evaluated in the field. Different measurements were taken to determine possible causes of clogging problems and to calculate the performance parameters of the irrigation systems. The measurements included, among others, the following:

- Delivery rate of emitters
- Water analysis
- Filter efficiency

A complete system evaluation was done according to the procedure described in ASAE EP 458,[4] where five dripper lines were evaluated at five locations. The delivery rates of 25 emitters per irrigation block were measured. Pressure readings were taken at the block inlets and along the manifolds. Apart from the $\bar{q}$ and CV_q values, the statistical discharge uniformity (U_s, %), which describes the uniformity of the emitter discharge in the block, was also calculated as shown as follows:

$$U_s = 100 - \text{CV}_q \tag{4}$$

A U_s value of 80% or higher is normally considered as an acceptable criterion.[4] The field emission uniformity (EU') was also used to judge the uniformity of emitter discharge within an irrigation block and is shown in eq 5. Table 4.3 reveals a comparison between U_s and EU as suggested for design purposes.

$$\text{EU}' = 100 \frac{q'_{min}}{\bar{q}} \tag{5}$$

where EU' is the field emission uniformity (%); q'_{min} is the measured mean of the lowest ¼ quarter of emitter discharge (L/h); and $\bar{q}$ is the measured mean emitter discharge (L/h).

4.3 RESULTS AND DISCUSSION

4.3.1 LABORATORY TESTS ON NEW DRIPPERS

Table 4.4 summarizes the results of the discharge and coefficient of discharge variation (CV$_q$) tests performed in the laboratory on emitters. The typical discharge categories for the turbulent dripper (A300 in this case) are presented in Fig. 4.3. For a turbulent pressure–compensated dripper (A200 in this case), typical discharge categories are presented in Fig. 4.4.

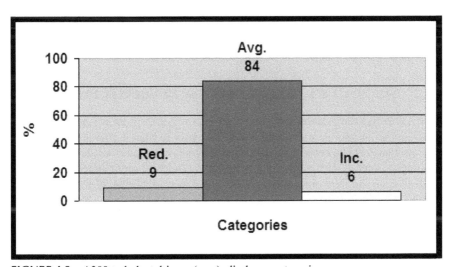

FIGURE 4.3 A300 turbulent dripper (new): discharge categories.

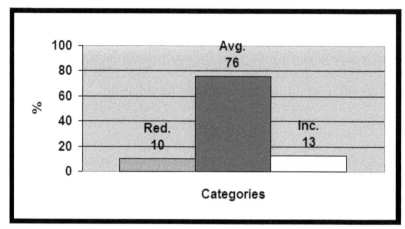

FIGURE 4.4   A200 turbulent dripper – pressure compensated (new): discharge categories.

Table 4.3 Comparison between U_s and EU for Design Purposes[4]

Classification	U_s (%)	EU (%)
Excellent	95–100	94–100
Good	85–90	81–87
Acceptable	75–80	68–75
Poor	65–70	56–62
Unacceptable	<60	<50

4.3.2   LABORATORY EVALUATION OF RETRIEVED EMITTERS

Table 4.5 shows the performance of the emitters retrieved from the field. The first set of dripper laterals were removed during February to March in year 1, 18 months after installation. The second set of dripper laterals were removed during April to May in year 2, after 30 months of use in the field. The third set of dripper lines were removed from the field during February to April in year 3, after 42 months of use. The final laboratory tests were concluded at the end of May in year 3. Table 4.6 summarizes all results of the treated and untreated emitters. A summary of the impact on emitters is shown in Table 4.7.

TABLE 4.4 New Emitter Test Results

Code of Dripper Tested	Sample Size Units	Dripper CV_q (%)	Average q, L/h	Percentage Emitters in Discharge Category						
				Discharge £Av-3CV	Discharge £Av - 2CV >Av - 3CV	Discharge £Av - 1CV >Av - 2CV	Discharge <Av + 1CV >Av - 1CV	Discharge ³Av+1CV <Av+2CV	Discharge ³Av + 2CV <Av + 3CV	Discharge ³Av + 3CV
A300	100	2.1	1.7	2	7	15	51	18	4	2
A500	100	4.2	2.2	10	9	11	37	14	12	6
A200	100	2.1	1.6	2	8	15	49	12	9	4
A100	100	2.4	1.23	2	2	15	61	14	5	1
A400	100	4.4	2.0	6	12	14	33	17	9	8

Note: Av: the average discharge q of the sample of 100 emitters (L/h); CV_q: coefficient of discharge variation of the sample (%).

TABLE 4.5 Summary of the Performance of Retrieved Dripper, as Tested in the Laboratory

	Year	Zone	Total Emitters Tested	Functional Emitters		Emitters Clogged by Root Intrusion		Emitters Clogged by Other Means	
				Number	%	Number	%	Number	%
Untreated	2002	4	215	176	81.9	21	9.8	18	8.4
		5	385	344	89.4	25	6.5	16	4.2
		6	275	272	98.9	0	0.0	3	1.1
		7	277	249	89.9	22	7.9	6	2.2
		8	123	102	82.9	5.0	4.1	16.0	13.0
	2003	4	237	159	67.1	44	18.6	34	14.3
		5	315	224	71.1	79	25.1	12	3.8
		6	266	254	95.5	6	2.3	6	2.3
		7	314	193	61.5	111	35.4	10	3.2
		8	185	133	71.9	48	25.9	4	2.2
	2004	4	160	95	59.4	44	27.5	21	13.1
		5	224	142	63.4	70	31.3	12	5.4
		6	244	228	93.4	15	6.1	1	0.4
		7	229	140	61.1	86	37.6	3	1.3
		8	205	112	54.6	90	43.9	3	1.5
Treated	2002	4	242	205	84.7	5	2.1	32	13.2
		5	483	459	95.0	9	1.9	15	3.1
		6	268	265	98.9	0	0.0	3	1.1
		7	233	215	92.3	10	4.3	8	3.4
		8	73	71	97.3	0	0.0	2	2.7
	2003	4	191	150	78.5	13	6.8	28	14.7
		5	–	–	–	–	–	–	–
		6	266	261	98.1	0	0.0	5	1.9
		7	262	197	75.2	45	17.2	20	7.6
		8	465	323	69.5	77	16.6	65	14.0
	2004	4	203	160	78.8	11	5.4	32	15.8
		5	244	192	78.7	15	6.1	37	15.2
		6	267	256	95.9	4	1.5	7	2.6
		7	219	176	80.4	38	17.4	5	2.3
		8	259	228	88.0	18	6.9	13	5.0

TABLE 4.6 Summary of All Results of the Treated and Untreated Emitters

Zone	% Functional				% Root Clogged				CVq (%)					Delivery Rate (L/h)				
	Untreated		Treated		Untreated		Treated		New	Untreated		Treated		New	Untreated		Treated	
	Year 1	Year 3	Year 1	Year 3	Year 1	Year 3	Year 1	Year 3		Year 1	Year 3	Year 1	Year 3		Year 1	Year 3	Year 1	Year 3
4	82	59	85	79	10	28	2	5	2.1	62.6	76.0	40.6	43.0	1.7	1.5	1.2	1.5	1.5
5	89	63	95	79	6	31	2	6	4.2	19.5	31.9	13.8	22.2	2.2	2.1	1.7	2.2	2.2
6	99	93	99	96	0	6	0	1	2.1	14.6	26.2	11.8	14.2	1.6	1.8	1.7	1.6	1.8
7	90	61	92	80	8	38	4	17	2.4	46.1	69.3	30.0	41.4	1.23	1.1	1.2	1.2	1.3
8	83	55	97	88	4	44	0	8	4.4	13.0	78.4	36.5	40.7	2.0	2.6	1.5	2.1	2.4
Average	87	66	94	84	5.6	29.4	2.6	7.4	3.04	31.16	56.36	26.54	32.3	1.75	1.82	1.46	1.72	1.84
% Decrease	24		11		—		—		—			—		—	2		—	
% Increase	—		—		425		184		—		81		22	—		—		7

TABLE 4.7 Summary of the Impact on Emitters (Comparing Year 3 Values)

Zone	% Functional		% Root Clogged		CVq %		Delivery Rate (L/h)	
	Untreated	Treated	Untreated	Treated	Untreated	Treated	Untreated	Treated
Average	66	84	29.3	7.5	56.36	32.30	1.46	1.84
% More	27						26	
% Less			74		43			

From Tables 4.6 and 4.7, we can conclude the following points:
- *Untreated blocks:* In the Treflan-treated blocks, the emitters showed 74% less root intrusion than the untreated blocks. The untreated blocks were 29.3% root intruded against the 7.5% of treated block, and there was also a significant increase in root intrusion of 425% over the 42-month period in the untreated blocks. The impact of clogging in the untreated blocks was evident with a 20% decrease in the emitter delivery rate and the worsening of the coefficient of variation (CV_q) from an excellent 3.04% to a very poor 56.36%. There was also an average of 34% of the emitters that were not functional after the 42-month testing period.
- *Treated blocks:* With the treated blocks, root intrusion could not be prevented completely, and after the 42-month period, 16% drippers were not functional (i.e., however, 27% better than the untreated blocks). The impact of the clogged drippers was that CV_q dropped from an excellent 3.04% to a poor 32.30%, which was still 43% better than the untreated blocks. However, the average emitter delivery rate stayed fairly constant.
- *Discharge categories:* Dripper lines with regular emitters showed a general tendency toward reduced average discharge due to partial or total clogging of drippers (Fig. 4.5).

On the other hand, drip lines with pressure-compensated emitters showed a general tendency toward increased discharge (Fig. 4.6).

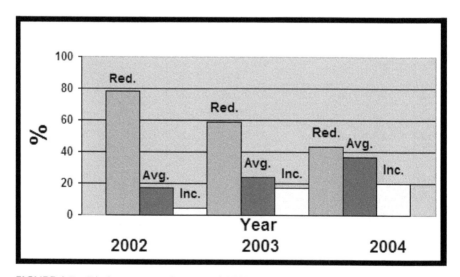

FIGURE 4.5   Discharge categories: treated A300 regular dripper.

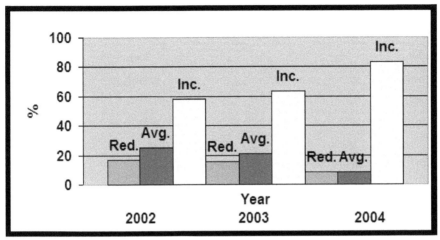

FIGURE 4.6   Discharge categories: treated A200 pressure-compensated dripper.

4.3.3   EVALUATIONS AT FARM SITE

Different measurements were taken to determine possible causes of clogging prob-
lems and to evaluate the performance parameters of the drip irrigation systems.
A summary of the water analysis is shown in Table 4.8 (Chemical) and Table 4.9
(Biological). Different emitter performance values are summarized in Table 4.10.

The cause of low values for Simu2 can be related to the water quality (high:
clogging potential) and the medium clogging potential due to the organic content of
the water and a low filtration efficiency.

TABLE 4.8   Chemical Analysis of Water

Site ID	pH	Iron (Fe) mg/L	Manganese (Mn) ppb	TDS mg/L	Danger of Clogging
Inyon.	6.95	0.39	65	8.0	Minor
Savan.	7.13	2.43	566	2050.0	Moderate
Simu1	7.35	0.65	74	13.6	Minor
Simu2	7.31	2.80	875	430.8	High

Note: ppb: parts per billion.

TABLE 4.9   Biological Analysis of Water

Site ID	Organic Chlorophyta	Clogging Potential
Inyon.	<20	Low
Savan.	+200	High
Simu1	<20	Low
Simu2	<100	Medium

TABLE 4.10   Values of Performance Parameters of Emitters

Site ID	EU %	EU_a %	U_s %	CV_q %	q_{max}	q_{min}	q_{ave}
Inyon.	90.9	89.1	90.9	9.1	1.3	1.0	1.1
Savan.	81.3	80.6	83.3	16.2	2.0	1.3	1.6
Simu1	93.8	89.0	83.1	16.9	1.9	1.5	1.6
Simu2	61.6	61.6	57.2	42.8	2.9	0	1.8

This chapter indicates that the use of root growth inhibitors is important and necessary for the effective functioning of drip lines and emitters. It was clear that emitter performance depends on many external and internal factors. In many cases, the performance is predictable, but there are also times when emitters can function totally independent of external factors. An in-depth knowledge and understanding of the emitters and their operation is then necessary to explain the difference in their behavior. In Table 4.11, an overall summary of all the relevant factors of emitter performance is shown to help to understand the operation and behavior of different emitters, which are available for subsurface drip irrigation (SDI). For new regular emitters, average coefficient of variation (CV_q) was 2.2% (excellent) compared with 5.2% (good) for the pressure-compensated emitters.

TABLE 4.11   Summary of Results for Emitter Characteristics and Performance Tests for Emitters

Emitter Type	Regular Emitters			Pressure-Compensated Emitters	
Code (Zone)	A300 (4)	A500 (5)	A200 (7)	A100 (6)	A400 (8)
Outlet aperture dimensions	1.8 mm diameter orifice	2 × 3.0 mm diameter orifice	13 mm slit	3.0 mm diameter orifice	2.0 mm diameter orifice

TABLE 4.11 *(Continued)*

Emitter Type	Regular Emitters			Pressure-Compensated Emitters	
Code (Zone)	A300 (4)	A500 (5)	A200 (7)	A100 (6)	A400 (8)
Inlet filter aperture dimensions (mm)	1.0×0.6	1.7×0.8	0.4×0.3	1.7×0.6	2.5×1.0
Number of apertures	22	5	370	4	4
Total inlet filter area (mm^2)	13	7	44	4	10
Flow path dimensions (mm)	0.7×0.7	1.0×0.9	1.0×0.3	1.0×0.8	1.0×1.0
Ratio[a]	1.43	1.89	1.33	2.13	2.50
Percentage clogging of the emitters by particles from inside the drip line (%)					
Laboratory clogging tests	79	45	10	88	91
Clogged field emitters	11.1	4.7	1.8	1.5	2.7
Percentage clogging of the emitters by roots only from outside the drip line (%)					
Untreated emitters	28	31	38	6	44
Treated emitters	5	8	17	1	8
Discharge category					
Untreated emitters	Reduced discharge	Reduced discharge	Increased discharge	Increased discharge	Reduced discharge
Treated emitters	Reduced discharge	Reduced discharge	Increased discharge	Increased discharge	Average discharge

[a]Ratio = LIFAD/SFPD, the *smaller* this ratio or number is, the *more effective* the inlet filter is, where LIFAD is the "Largest" inlet filter aperture dimension and SFPD is the "Smallest" flow path dimension.

4.4 CONCLUSIONS

Drip irrigation is considered as the most efficient irrigation system. However, based on past research and this study, it was found that drip irrigation system can also be inefficient as a result of water quality, root intrusion mismanagement, emitter clogging, and maintenance problems. The use of root growth inhibitors in the SDI system should be done according to the relevant recommendations from the suppliers of the chemicals. This is to establish the registration (legal approval) of the specific product for use on the specific crop in SDI applications.

At the experimental site and for the Treflan-treated blocks, the emitters showed 74% less root intrusion than the untreated blocks. The untreated blocks were 29.4% root intruded against the 7.5% of treated blocks, and there was also a significant increase in root intrusion of 425% over the 42-month period in the untreated blocks. The impact of clogging in the untreated blocks was evident with a 20% decrease in emitter delivery rate and the worsening of the CV from an excellent 3.04% to a very poor 56.36%. There was also an average of 34% of the drippers that were not functional after the 42-month testing period. With the treated blocks, root intrusion could not be prevented completely, and after the 42-month period, 16% drippers were not functional (i.e., however, 27% better than the untreated blocks). The impact of the clogged drippers was that the CV dropped from an excellent 3.04% to a poor 32.30%. However, average emitter delivery rate stayed fairly constant.

For the farm site evaluations, the CV varied from a marginal 9.1% to a poor 42.8%. The emission uniformity (EU_a) varied from a good 89.1% to an unacceptable 61.6%. The research concluded that the practice of subsurface drip must be done with caution and that good management and maintenance practices with the chemical treatment of the soil surrounding the dripper lines are of utmost importance.

4.5 SUMMARY

Clogging of the emitters is one of the most serious problems associated with SDI. Various approaches in preventing the clogging of emitters include filtration, flushing, chemical treatment of the irrigation water and of the soil surrounding the drip lines, as well as the chemical treatment of the lateral polymers.

This research was carried out at the ARC-IAE, South Africa. The blockage potential due to root intrusion was evaluated for five types of emitters supplied by three drip irrigation companies. Research was carried out at an experimental field on five different types of emitters, and retrieved dripper lines from the experimental field were tested in the laboratory. Evaluations were also carried out in field under farming conditions.

With new regular emitters, average coefficient of variation (CV_q) was an excellent 2.2% and good 3.2% for the pressure-compensated emitters. For the experimental site and the Treflan-treated blocks, the emitters showed 74% less root intrusion than the untreated blocks. The untreated blocks were 29.4% root intruded against the 7.5% of treated blocks, and there was also a significant increase in root intrusion of 425% over the 42-month period in the untreated blocks. The impact of clogging in the untreated blocks was evident with the 20% decrease in emitter delivery rate and the worsening of the CV from an excellent 3.04% to a very poor 56.36%. There was also an average of 34% of the drippers that were not functional after the 42-month testing period.

With the treated blocks, root intrusion could not be prevented completely, and after the 42-month period, 16% drippers were not functional (i.e., however, 27% better than the untreated blocks). The impact of the clogged drippers was that the CV dropped from an excellent 3.04% to a poor 32.30%. However, average emitter delivery rate stayed fairly constant. At the farm site evaluations, the coefficient (CV) varied from a marginal 9.1% to a poor 42.8%. The emission uniformity (EU$_a$) varied from a good 89.1% to an unacceptable 61.6%.

The research concluded that the practice of subsurface drip must be done with caution and that good management and maintenance practices with the chemical treatment of the soil surrounding the dripper lines are of utmost importance.

KEYWORDS

- Agricultural Research Council, ARC
- Agricultural Research Council – Institute for Agricultural Engineering, ARC-IAE
- ASAE
- chemical treatment
- clogging
- coefficient of variation
- delivery rate
- drip irrigation
- drippers
- emission uniformity
- emitter
- emitter clogging
- field evaluation
- filtration
- flushing
- Institute for Agricultural Engineering, IAE
- pressure-compensated emitters
- retrieved dripper lines
- South Africa
- subsurface drip irrigation
- Treflan
- USA

REFERENCES

1. Koegelenberg, F. H.; Reinders, F. B.; Van Niekerk, A. S.; Van Niekerk, R.; Uys, W. J. Performance of Surface Drip Irrigation Systems Under Field Conditions; WRC Report No. 1036/1/02; Water Research Commission, 2003. ISBN No. 1-86845-973-X.
2. Reinders, F. B.; Smal, H. S.; Van Niekerk, A. S.; Bunton, S.; Mdaka, B. Sub-Surface Drip Irrigation: Factors Affecting the Efficiency and Maintenance; WRC Report No K5/1189/4; Water Research Commission.
3. ISO. *ISO/TC 23/SC 18 N 89. Irrigation Equipment: Emitters Specifications and Test Methods*; International Standards Organization (ISO), 1983.
4. ASAE. *EP 458: Field Evaluation of Micro-Irrigation Systems*; ASAE: USA, 1997.

CHAPTER 5

PERFORMANCE CHARACTERISTICS OF MICRO SPRINKLER

V. G. NIMKALE, S. R. BHAKAR, H. K. MITTAL, and B. UPADHYAY

CONTENTS

Edited and abbreviated version of "V. G. Nimkale, 2014. Evaluation of micro sprinkler characteristics under various operating conditions. M. Tech. Thesis for Department of Soil and Water Engineering, CTAE, Udaipur – India," Email: vinodnimkale@gmail.com.

5.1 INTRODUCTION

Water is the elixir of life. It is a vital element that significantly affects all aspects of our daily life. Water is an invaluable natural resource, a basic human need, and the country's treasure. It is construed as a soul for "life-sustaining profession" and the economic development of any country. Water is one of the most important inputs among all that required for biological activities of an agricultural plant. Water is a scare resource, and its demand is looming unintermittently large. The water table already touches its fathom depth in many parts of the world. Therefore, efficient use of available water is extremely important.

In India, the annual average precipitation is 1140 mm, and about 90% of water is used for agriculture. The country has 145 million hectare net sown area out of a total 175 million hectare gross cropped area. As per the estimate, the total irrigated area is 46 million hectare. The population of India has already crossed 1 billion and is likely to stabilize at 1.64 billion by AD 2050. Hence, the country's policy makers will have to plan for increasing the food grain production from the current level of 200 to 450–500 million tons by 2050.[1] Hence, efforts are exigently needed to maximize the crop production per unit of water used and area under irrigation.

Agriculture is the backbone of Indian economy as it ensures food security, engenders employment, helps to palliate poverty, and contributes significantly to the export of the country. It contributes nearly 19% of gross domestic product, and about 68% of the Indian population is reliant on agriculture for their livelihood.[2] Irrigation is an important input, which has strategic impact on agricultural production. In India, rainfall is very erratic, and it is not sufficient to fulfill the water requirement of crops for irrigation, in many areas. In order to increase irrigation potential, it is utmost important to make the best use of available water for irrigation by the way of adopting efficient irrigation water management techniques or practices.

Therefore, if the world food crisis is to be solved, there does not seem to be other alternative, but to increase the total area under irrigation. This is possible only through better water management practices in the field and by introducing advance methods of irrigation wherever possible, compatible with socioeconomic conditions of the area.

5.1.1 IRRIGATION METHODS

Irrigation is defined as the artificial application of water to soil for the purpose of supplying the moisture beneficial or essential to the plant growth. Irrigation water may be applied to the crop by (i) conventional surface irrigation methods (i.e., border irrigation, furrow irrigation, check basin irrigation, etc.); (ii) subsurface irrigation method (i.e., open ditches and itile drain); and (iii) pressurized irrigation methods (sprinkler irrigation and micro or drip or trickle irrigation). Water supply, type of soil, topography of the land, and crops to be irrigated determine the correct method of irrigation to be used. Whatever is the method of irrigation, it is necessary to design the system for the most efficient use of water by the crop. Most farmers in developing countries as well as in developed countries still use gravity flow system to irrigate the fields. The age-old irrigation method distributes water through unlined field channels. In a large number of cases, only a small portion (about 30–35%) reaches the field, and half of the water is not used by the plant because of deep percolation and evaporation losses. Therefore, it is necessary to shift from traditional low-efficiency irrigation methods to high-efficiency irrigation methods such as micro sprinkler irrigation for the efficient utilization of scarce water resources to save energy.

5.1.2 MICRO SPRINKLER IRRIGATION

Micro sprinkler irrigation is a method of micro irrigation. Micro sprinkler is a low-volume sprinkler (i.e., 30–300 lph) that operates at low pressure (i.e., 1.0–2.5 kg/cm²). It applies water in the form of a spray. Micro sprinkler irrigation method is the intermediate irrigation system between sprinkler irrigation and drip irrigation. It requires lesser energy than conventional sprinkler system and is lesser susceptible to clogging than emitters. It has an area of coverage much larger than emitter, but much lower than the conventional sprinkler system. It has low cost of installation than drip system because the number of lateral and outlets is reduced.[3] Also in micro sprinkler irrigation, the root system is developed evenly and spread densely throughout wetted soil volume. This ensures better supply of water and nutrient to plant. Micro sprinkler system has a wide range of uses as in fertilizer application, herbicide application, and frost protection, cooling of greenhouses, and poultry houses. This system can be run either continuously or intermittently at a desired rate of application.

5.1.3 OPERATING CHARACTERISTICS OF MICRO SPRINKLER IRRIGATION SYSTEM

Irrigation uniformity, uniformity coefficient (CUC), distribution uniformity (DU), wetted diameter, water application rate, pressure–discharge relationship, and precipitation characteristics are major features of micro sprinkler system, which affect the design and the operational efficiency of this system.

5.1.3.1 IRRIGATION UNIFORMITY OF A MICRO SPRINKLER

Irrigation uniformity is related to crop yields through agronomic effects of under- or overwatering. Insufficient water leads to high soil moisture tension, plant stress, and reduced crop yields. Excess water may also reduce crop yields below potential levels through mechanisms such as leaching of plant nutrients, increased disease incidence, or failure to stimulate the growth of commercially valuable parts of the plant.

Because irrigation uniformity relates to crop yield and the efficient use of resources, engineers regard it as an important factor to be considered in the selection, design, and management of micro sprinkler irrigation. Various measures of uniformity are used as indices of performance by which sprinklers and sprinkler spacing are judged, and they may also be used to set hydraulic limitations on the micro sprinkler pipe network. Irrigation uniformity is a key component in overall irrigation efficiency and hence plays an important role in the irrigation scheduling to meet crop water requirements.

5.1.3.2   UNIFORMITY COEFFICIENT

This is another parameter developed by Christiansen.[4] which is widely used to evaluate micro sprinkler irrigation uniformity. A measurable index of degree of uniformity obtained from any size of sprinkler operating under given condition is known as uUC. A UC of about 85% or more is considered to be satisfactory.[3] The data on UC are basis for selecting the combinations of spacing, discharge, nozzle size, and operating pressure for obtaining higher irrigation efficiency at specific operating conditions.

5.1.3.3   DISTRIBUTION UNIFORMITY

The DU is a measure to know how evenly water is applied across a field during irrigation. For example, if one unit of water is applied in one part of the field and only half unit is applied in another part of the field, this is a poor DU. The DU is expressed as a percentage between 0 and 100. However, it is impossible to attain 100% in practice. If DU lesser than 70% is considered to be poor, DU between 70 and 90% is good, and that greater than 90% is excellent. In short, poor DU means either too much water has been applied, which is costing unnecessary expenses, or too little water is applied causing water stress to crops. Therefore, micro sprinkler system must have a good DU for better irrigation efficiency.

5.1.3.4   WETTED DIAMETER OF A MICRO SPRINKLER

The installation of system is controlled by wetted diameter. Wetted diameter of a micro sprinkler or nozzle is the distance between the micro sprinkler and dry soil on both sides of the sprinkler. The wetted diameter is controlled by operating pressure of the device, the trajectory angle, the water leaving the device, and the location where the device is mounted relative to the soil surface. Hence, for any type of micro sprinkler or nozzle, a range of wetted diameter is possible.

5.1.3.5   WATER APPLICATION RATE OF A MICRO SPRINKLER

The rate at which a micro sprinkler system irrigates the soil and when a group of them is operating close together is called the application rate. The water application rate by micro sprinklers is limited by the infiltration capacity of the soil. The infiltration capacity of the soil is the maximum rate at which water can enter the soil at a given time. Application with higher rates than infiltration capacity of the soil results in runoff, accompanied with poor distribution of water, loss of water, and soil erosion. Normally, micro sprinkler irrigation systems are designed in such a way that no runoff occurs. Thus, the rate at which a micro sprinkler is designed to apply water is less than the infiltration capacity.

5.1.3.6   PRESSURE–DISCHARGE RELATIONSHIP FOR MICRO SPRINKLER

The spray distribution characteristics (DC) of micro sprinkler heads are typical of operating pressure. At low pressures, the drops are larger, and water from the nozzles falls in a ring away from the micro sprinkler. Contrary to this, at high pressures, the water from the nozzle breaks up into very fine drops and fall very near to the micro sprinkler.

5.1.3.7   PRECIPITATION CHARACTERISTICS

The design of micro sprinkler irrigation system depends on precipitation characteristics (effective radius, average application depth, effective maximum depth, absolute maximum depth, mean application depth), DC, coefficient of variation (CV), etc.

5.1.4   DESIGN CONSIDERATIONS

Technology in irrigation like other fields is changing very rapidly. The worldwide research in irrigation is being carried out to use the water judiciously and to obtain more crop yield per unit volume of water; to recommend the most suitable methods of irrigation in varying land topography, soil depth, and macro climate; and to assess feasibility of the method to meet the food requirements of ever-increasing population. Micro sprinkler has not been as popular as drip irrigation system because of several factors, such as lack of information about water requirement for orchard crops, design skills and high investment cost, and lack of detailed and generalized specification system.

The ideal micro sprinkler irrigation system assures delivery of equal volume of water for all micro sprinklers. But practically, it is difficult to achieve this as the flow from the micro sprinkler is affected by variation in water pressure in the distribution system, which ultimately affects the wetting diameter. The water network flows at different operating pressures may be determined by pressure–discharge relationships.

Uniform water distribution by any irrigation system maximizes crop yield and improves the quality of produce. In shallow-rooted crops, higher uniformity of application is desirable, whereas in deep-rooted crops, a lower uniformity of application may be tolerable. The uniform water distribution is also necessary for the efficient use of available irrigation water. For this reason, the emission uniformity (EU) and UC may always be taken as design variables for micro sprinkler irrigation system.

For having an acceptable irrigation pattern, the information regarding wetting diameter at different operating pressures is required because it determines the optimum overlap to be provided when the system is operated in actual field conditions. The specific design requirements of a micro sprinkler system involve the optimum spacing of emitting devices and water application rates.

The above discussion suggests that the knowledge of pressure–discharge relationships, precipitation pattern, wetting diameter, UC, EU, and DU is essential for optimum design and operation of micro sprinkler irrigation system. Considering this point in view, this chapter discusses the performance characteristics of a micro sprinkler with the following objectives:

1. To determine the precipitation distribution pattern of micro sprinkler.
2. To determine the application uniformity characteristics of micro sprinkler as influenced by the operating pressure and the micro sprinkler spacing.
3. To develop the pressure–discharge relationships for a micro sprinkler.

5.2   REVIEW OF LITERATURE

5.2.1   PRESSURE–DISCHARGE RELATIONSHIPS

Seginer[5] showed that the operating pressure of the micro sprinkler irrigation is the most important factor. Higher pressure causes longer radius of wetting, fine drops, and more even distribution of water on the ground. Keller and Karmeli[6] described the pressure–discharge relationship for emitting device as follows:

$$Q = K (H^x) \tag{1}$$

where Q is the emitter discharge, lph; K is the constant of proportionality that characterizes the emitter; H is the operating pressure head, m; and x is the exponent that characterizes the flow regime. Value of x characterizes the flow regime of emitter. For a fully turbulent flow, $x = 0.5$; for a partially turbulent flow, $0.5 < x < 0.8$; for unsteady flow regime, $0.8 < x < 1.0$; and for laminar flow, $x = 1.0$. The long path emitters have exponent from 0.6 to 1.0. Some emitters provide varying degrees of flow regulation, and "x" may be less than 0.5. Considerable regulation is achieved with "x" ranging from 0.3 to 0.4.

Voigt[7] described the relationship between the hydraulic performance of rotary sprinklers that was expressed in terms of discharge coefficient as a measure of range and design with respect to the cone angle of a nozzle and sprinkler head dimensions. He concluded that the width of sprinkler head should be about three times than that of the nozzle. Giari et al.[8] tested five drippers and five micro jet sprayers at a pressure range of 0.4–1.5 kg/cm². They found considerable variations in discharge rates and the uniformity of coverage even for emitters of the same type. They also observed one micro jet affected by the position of the flute mouthpiece outlet in relation to the direction of the water flow. Singh et al.[9] carried out an experiment on the performance evaluation of micro sprinklers and suggested the constraints for its adoption. They also evaluated the system by estimating flow variation in lateral lines and field EU. They developed the following pressure–discharge relationship based on actual field data:

$$Q = 62.25 \ (P)^{0.58} \tag{2}$$

where Q is the micro sprinkler discharge, lph, and P is the operating pressure, kg/cm².

Firake et al.[10] carried out the evaluation of hydraulic performance of micro sprinklers. They showed that the micro sprinkler discharge was increased with an increase in pressure. They considered the effective wetted area of soil for the calculation of the UC of micro sprinklers. Sakore[11] studied pressure–discharge relationship for three types of micro sprinklers. He found discharge in the range of 44–63 lph and observed an increase in discharge with operating pressure, whereas a decrease in discharge was observed with increase in stake height.

Firake and Salunkhe[12] reported that the micro sprinkler discharge was increased with an increase in pressure. They further commented that the effective wetted area of the soil should be considered only for the calculation of the UC of micro sprinkler. They also found that downward vertical movement of water in the soil was increased with increase in operating pressure. Shinde et al.[13] studied the pressure–discharge relationship of static micro sprinklers. They showed that the discharge was increased by 63.5% with an increase in pressure head by 166%. They also reported that the average EU of the static micro sprinkler system was 91.0%.

Singh[14] carried out the performance evaluation of three different models of micro sprinklers as influenced by design variables. He analyzed the performance of a particular type of micro sprinkler by considering the pressure–discharge relationship. He found an increase in discharge with an increase in pressure. He also observed that the observed discharges were in conformity with that supplied by the manufacturer. He developed a power form relationship between the pressure and the discharge.

Gawali and Budhan[15] conducted the study on pressure–discharge relationship, UC, wetted throw, and wetting pattern for five micro sprinklers for operating pressures of 0.5, 1.0, and 1.5 kg/cm² and stake inclination angles of 90, 80°, 70°, and 60° with respect to the horizontal. The discharge was observed in the range of 20.54–64.09 lph and was increased with operating pressure, whereas it was not influenced with change in inclination angle. The UC values ranged from 9.3 to 55.5%. The UC values were reduced with increase in operating pressure, whereas these were not influenced with the inclination angle. However, the inclination angle influenced the wetting pattern. For micro sprinkler (M_2), the wetting pattern shifted toward the direction of inclined ride, whereas other four micro sprinklers showed contraction in shape as compared with vertical position. M_3 sprinkler showed spot application of water. The precipitation depth collected on inclined side for all five sprinklers was more than the opposite side. The maximum throw observed for M_1, M_2, M_3, M_4, and M_5 micro sprinklers were 2.8, 3.0, 5.5, 5.4, and 5.2 m, respectively. Data on precipitation pattern by Lonkar and Dhage[16] revealed superior precipitation pattern for an operating pressure of 1.3 kg/cm². The value of the manufacturer's CV was 0.12, which was less than the acceptable limit of 0.15 for the category of other types of micro irrigation methods. Lonkar and Dhage[16] developed pressure–discharge relationships for two types of micro sprinklers designated:

$$Q = 22.39 \ H^{0.47}, r = 0.82 \ \text{(for micro sprinkler A)} \tag{3}$$

$$Q = 21.87 \ H^{0.48}, r = 0.98 \ \text{(for micro sprinkler B)} \tag{4}$$

where Q is the discharge, lph, and H is the operating pressure, kg/cm².

Suryawanshi et al.[17] studied pressure–discharge relationship, precipitation pattern, and performance parameters of a micro sprinkler. They found that discharge was a function of operating pressure head. Uniformity precipitation pattern was increased with increased pressure head up to a certain limit. Greater precision was required in the manufacturing of micro sprinklers. They further reported that the micro sprinklers are suitable for under tree irrigation. Singh et al.[18] studied pressure–discharge relationship, EU, coefficient of manufacturing variation, wetting diameter, and uniformity of application of a micro jet. These parameters for micro jet were determined for different combinations of pressures ranging from 0.5 to 1.7 kg/cm² and stake heights of 0–10 cm. The estimated value of nozzle exponent was found in the recommended range. Based on the value of nozzle exponent, the micro jet was classified as non-pressure compensating. The EU was more than 90%. The micro jet was classified as excellent on the basis of value of coefficient of manufacturing variation; the wetting diameter was increased with increase in operating pressure and stake height. The UC, DU, and DC were increased with increase in operating pressure and stake height.

Pampattiwar et al.[19] evaluated pressure–discharge relationship, manufacturing CV, and precipitation pattern for micro sprinkler. They found that discharge was increased with an increase in operating pressure. More precision was observed during the manufacturing process of SP-1- and SP-2-type micro sprinklers, whereas poor manufacturing quality was observed in SP-1-type micro sprinkler and flat elliptical distribution profile. Minimum CV in SP-2 type of micro sprinkler indicated its superiority over SP-1 and SP-2 type of micro sprinklers. Patil et al.[20] studied pressure–discharge relationship, manufacturer's CV, and precipitation pattern at the operating pressure heads of 1.0–2.4 kg/cm². Precipitation characteristics were lower for S-3 type of micro sprinkler compared with other two types of micro sprinklers under study, indicating the superiority of S-3-type micro sprinklers.

Barragan and Wu[21] found that simple pressure parameters along a lateral line or in a rectangular sub-main unit (maximum pressure, minimum pressure, and average pressure) can be used for the hydraulic design of micro irrigation systems. This was based on the fact that simple pressure ratios (minimum pressure to maximum pressure and minimum pressure to average pressure) are all indications of the uniformity of micro irrigation systems. A definite relationship between the total friction pressure loss and maximum and minimum pressure difference, or average and minimum can be determined for a micro irrigation system under different field slope situations. When a nominal pressure head (10 m) is set for the average or maximum pressure, the minimum pressure can be determined based on the selected design criteria. The total friction pressure loss can be considered as the sum of the total friction pressure loss for the lateral and sub-main. The length of the lateral and the size of the sub-main can be determined from the respective total friction pressure losses.

For rotary sprinkler irrigation system, Jitander[22] developed pressure–discharge relationship as described by Keller and Karmeli[6] of the form of $Q = [KP^x]$. Discharge was measured with the help of flexible tubes connected to nozzle mouth. The value of discharge exponent (x) ranged from 0.41 to 0.566.

5.2.2 PRECIPITATION PATTERN

Fisher and Wallender[23] studied the influence of collector diameter and test duration on uncertainly in depth of water caught. The CV of water application was decreased as collector size and test duration were increased. They concluded that for larger collectors, fewer tests were needed to reach a stable CV. Thus, they recommended larger collectors to maintain accuracy and reduce test duration and number of tests. Madramootoo et al.[24] tested five online orifice emitters at pressure ranging from 69 to 138 kPa. They concluded that the coefficient of manufacturing variation was not affected by pressure in case of pressure-compensating emitters, but it was affected by the operating pressure for non-pressure- compensating emitters.

Boman[25] concluded research on several micro irrigation spinner and spray emitters to evaluate the distribution patterns and relationship between the operating pressure and the discharge. Emitter flow rate pattern and DU were measured for each type of emitter, which had higher uniformity of water application than spray type under no wind conditions. Most spray emitters had 50–75% of the wetted area receiving insignificant water application, whereas 10–15% of wetted area received more than three times the average application. Spinner emitter, however, had 30–80% of coverage. None of the spinner models had areas of application greater than four times the mean application depth of emitter.

Gutal et al.[26] carried out a comparative study of drip, micro sprinkler, and bi-wall and border irrigation. They reported that the uniformity of micro sprinkler with 1.2 × 1.2 m spacing and 30 cm riser height was 76.75% at 1.0 kg/cm[2] pressure head. It was observed that the precipitation rate was decreased with an increase in riser height. They also reported that maximum wetted diameter of soil was 3.1 m at 30 cm riser height with a precipitation rate of 4.6 cm/h. Pathare[27] indicated that the wetting pattern of micro sprinkler for all operating pressures showed a triangular shape. The discharge was in the range of 24–34.2 lph for operating pressure in the range of 1.2–1.8 kg/cm[2]. The resulting wetting pattern was obtained due to overlapping of micro sprinkler when spaced at 1.0 × 1.0 m to 2.5 × 2.5 m. He indicated that at lower spacing and lower operating pressures, nonuniformity of water application was minimized and thus resulting more uniform depth of precipitation.

Aragade and Thombal[28] found that discharge was increased with operating pressure for both Black-30 and violet-45 micro sprinklers. They also determined spray patterns for different micro sprinklers. From spray pattern, maximum and minimum diameters of throw were determined. They were approximately same as given by the manufacturers in the technical manual. Pandey et al.[29] quoted that the maximum radii of throw obtained for A, B, C, D, and E micro sprinklers were 4.98, 3.94, 3.93, 2.96, and 2.23 m, respectively. The effective radius was lower for all the micro sprinklers using a Keller method[30] when compared with the Boman method.[25] According to the Keller criteria of DC, micro sprinklers B and D were found to be satisfactory. The DC values obtained by the Boman method were higher than the Keller method for all the micro sprinklers. Both the methods for the determination of DC value are arbitrary in providing criteria for the evaluation of water distribution of micro sprinklers. Further, higher precipitation rate was obtained by Boman method than those obtained by Keller method.

Shete and Modi[31] performed experiments by altering nozzle sizes, operating pressure, and riser high to investigate how water distribution around sprinklers was affected by the layout pattern of the catch cans. In the trials, the catch cans were arranged around a sprinkler in a grid of 2 × 2 m spacing and along a radial line at 1 m center. Grid spacings of 6 × 6 m, 6 × 12 m, 6 × 18 m, and 12 × 12 m were simulated. Such analysis gave conservation estimates of UC and DU. Nozzle size was the dominant parameter controlling water DU.

Vishnu and Santhana[32] studied performance of four spinner emitters at pressures of 49.35, 98.07, and 147.10 kPa to study the distribution pattern and uniformity of water application. Emitters were positioned on stakes 0.20 m above the top of the catch cans, which were placed at 0.60 m grid intervals in a matrix. The distribution pattern was obtained by plotting the depth of water collected at differences from the emitters as a percentage of the average application depth. The DC, CV and ratio of effective maximum depth (Dxe) to average application depth (da) were evaluated to study the DU of emitters. Generally, spinner emitters had higher DU than spray emitters under no wind conditions.

Mateos[33] developed a model for simulating precipitation from single sprinklers. Results indicated that operating pressure determined sprinkler flow and maximum throw. Wind and evaporation distorted the distribution patterns. Application of the model showed the impact of system management and design, field topography and wind on irrigation uniformity. Management factors such as lateral operation time or riser inclination may account for a large part of the field precipitation variations. A rough topography may also reduce uniformity significantly. Wind speed is important when it exceeds 1.8–2.0 m/s.

Singh et al.[34] determined the UC on the overlap area by four micro sprinklers of nozzle size 1.12 mm at different operating pressures (i.e., 0.2–1.2 kg/cm² with increasing 0.2 kg/cm² pressure) and nozzle spacings (i.e., 21.5 cm × 21.5 cm, 36 cm × 36 cm, 40 cm × 40 cm, 36.6 cm × 36.6 cm, 48 cm × 48 cm, and 41.6 cm 41.6 cm). They concluded that (i) the application profiles of micro sprinklers are not of continuous type, and only 50% of the radius of throw received water. (ii) The spacing of micro sprinklers along the lateral or the spacing of lateral along the main and sub-main should be taken equal to the radius of throw. (iii) Operating pressure and spacing for different makes, models, and nozzle sizes should be recommended for maximum UC.

DeBoer[35] found that the water application distribution patterns for many field irrigation data sets followed a trapezoidal pattern. The trapezoidal shape influenced the magnitude of potential surface runoff. There were scenarios in which the trapezoidal pattern produced more estimated potential runoff than the commonly used elliptical pattern and other scenarios in which it yielded less potential runoff. Each of the rising and falling segments of impact sprinkler patterns ranged from 10 to 20% of total application time, where the remainder of the application time had a relatively constant application rate.

Faci et al.[36] reported that the wetted width produced by the rotating spray plate sprinkler (RSPS) was larger than that of the fixed spray plate sprinkler (FSPS), for nozzle diameters of 6.7 and 7.9 mm. Also the peak instantaneous precipitation rate of the RSPS was smaller than that of the FSPS.

In a laboratory experiment on RSPS, DeBoer[35,37] concluded that maximum wetted radii were positively related to increased sprinkler nozzle size and nozzle pressure. Nozzle diameter had a minimal effect on drop size, but nozzle pressure had significant inverse influence. Patil et al.[20] found that precipitation characteristics were lower for S-3 type of micro sprinklers compared with other two types of micro sprinklers under study, indicating the superiority of S-3-type micro sprinklers.

Clark et al.[38] conducted field measurement experiments on large-scale sprinkler irrigation systems with fixed-plate (FP) low drift nozzle (LDN). Scenarios included sprinkler operating pressures of 416, 104, and 138 kPa; sprinkler spacing 1.83, 2.44, 3.05, and 3.66 m; and nozzle orifice sizes of 4.76–7.94 mm with a flow range of 0.16–0.77 lps. Simulated patterns and UC values compared well with field-measured patterns and UC values for the respective sprinkler size, spacing, and operating pressure combinations. UC values from simulated patterns were highest for closer sprinkler spacing scenarios (<2.4 m) and higher operating pressures (104 and 138 kPa; still in the low range for sprinkler systems). However, evaporation and wind drift losses could be higher than that with the lower operating pressures, thus reducing the overall application efficiency. Based on the spacing, nozzle size, and operating pressure scenarios tested in this research, sprinkler spacing to wetted diameter ratios should not exceed 0.20 in order to achieve UC in excess of 90 under no-wind conditions with FP, LDN-type sprinklers.

Dogan et al.[39] concluded that the collectors with a 10 cm opening may not provide reliable irrigation depth data under FP sprinkler packages that produce distinct streams of water. Collectors with 15 cm diameter openings provided acceptable results under the FP sprinkler packages. The 10 cm opening collectors accurately measured both irrigation depth and uniformity under spinning plate and wobbling-plate sprinkler irrigation packages that produced smaller, more evenly distributed irrigation droplets with no distinct streams or jets.

Nehete et al.[40] found that precipitation pattern and precipitation distribution profile (average application depth, effective maximum depth, absolute depth, mean depth, DC, and CV) indicate that SP-2 type of micro sprinkler has flat elliptical distribution profile, and precipitation contours are comparatively at equidistance minimum variation in average application depth, effective maximum depth, absolute maximum depth, and mean depth together with maximum value of DC and CV for SP-2 type of micro sprinkler operated at 2.5 kg/cm^2 pressure.

Sourell et al.[41] reported that the simulated UC of a lateral equipped with RSPS at 3 and 4 m overlapping distance in different working conditions was always higher than 87%, and it has an average of 91.80%. The water application found for the analyzed RSPS was trapezoidal, resulting in higher constancy.

Playan et al.[42] reported that water application from FSPS often resulted in a bimodal pattern, whereas RSPS produced bull-shaped or triangular pattern. At a nozzle elevation of 2.4 m and an operating pressure of 140 kPa, the wetted diameter was 1.6 m larger for the RSPS than for the FSPS.

James et al.[43] studied a method for the determination of optimum design parameters of a micro sprinkler, in closed area with four micro sprinklers from four manufacturers named as A, B, C, and D. The independent parameters such as input pressure (P) and stake height were varied, and dependent parameters such as discharge (q, m³/s), effective radius (m), average depth of water (d), and UC of a single emitter (UC, %) were evaluated. The data of these dependent variables of the four types of emitters were subjected to optimization by multi-objective programming following three-dimensional surface plot technique. For this purpose, the software goal-attaining function of MATLAB6 was used. Finally, the optimum values of dependent variables were obtained, which were used in the design of the network of the micro sprinkler irrigation system. Poul et al.[44] studied hydraulics of micro sprinkler and drip irrigation (discharge, radius of coverage, and UC at different operating pressures), and geometrical patterns for higher UC. From hydraulics, it was concluded that micro sprinkler gave best radius of coverage and discharge suitable for tuberose crop with square (2.4 × 2.4 m) geometrical pattern.

Kadam et al.[45] studied hydraulic performance of micro sprinklers under laboratory conditions in terms of manufacturer's CVs and precipitation pattern for the operating heads in the range 1.0–2.4 kg/cm². Manufacturer's CV was estimated. Micro sprinklers SP-2 and SP-3 recorded values in the range of 0.02–0.06 and 0.015–0.034, respectively, whereas micro sprinkler SP-1 recorded the value in the range of 0.28–0.41 indicating poor manufacturing quality. Precipitation pattern was studied for all three micro sprinklers under study according to the procedure suggested by Keller[46] in terms of distribution profile, effective radius, average application depths, effective maximum depth, absolute maximum depth, mean depth, DC, and CV. The values of precipitation patterns indicated the superiority of SP-2 type of micro sprinkler over other two types of micro sprinklers under study.

5.2.3 CHARACTERISTICS OF APPLICATION UNIFORMITY

The application uniformity can be evaluated in terms of EU, DU, and UC. These parameters are reviewed in this section.

Keller and Karmeli[6] described EU as a relationship between minimum and average emitter discharge rates within the system. They found that this relationship is the most important factor in uniformity of application because the primary objective of irrigation system is to ensure enough system capacity to adequately irrigate the least watered area. They described the relationship as follows:

$$EU = 100 \times \frac{q_n}{q_a} \tag{5}$$

where q_n is the average of lowest quarter of emission point discharge for field data, lph; and q_a is the average discharge of test sample operated at reference pressure head, lph.

Solomon[47] showed that uniformity test results varied significantly, even under similar test conditions. The variation was correlated with the UC value itself, and the statistical significance of such variation was explored. Forkel and Mirshei[48] described a formula for calculating sprinkling range as a function of operating pressure, nozzle diameter, and a constant K, which is specific for a particular sprinkler. K is an indicator of the spraying power of a sprinkler. They also developed a nomograph between operating pressure, nozzle diameter, and sprinkling range. Ricardo et al.[49] found that zero leaching can be accomplished with a sprinkler system by specifying DU. Uniformity can be set to assume that minimum applications of water will keep plants alive without over-irrigating to cause leaching anywhere in the field.

Post et al.[50] indicated that all low-flow sprinklers demonstrated poor rating regardless of the pattern or application uniformity procedure. They recommended that the catch cans should be placed in a square matrix pattern over one quarter of throw area and perpendicular to the low-flow sprinkler arm. Further, they found that many low-flow sprinkler produced "doughnut" patterns. Hill et al.[51] found that a sinusoidal oscillation at 0–103 kPa for all emitters over 1-min cycle provided an equivalent and better Christiansen's uniformity coefficient (CUC) than that when operated at a steady pressure of 207 kPa. They showed that the EU for a pressure of 228 kPa steady, 103 kPa steady, and 0–103 kPa oscillating were 92.2, 91.5, and 91%, respectively.

Gutal et al.[52] observed the optimum wetted soil diameter when the system was operated at 1.5 kg/cm² pressure with a spacing of 3 × 3 m, application rate of 3.3 mm/h, and a stake height of 30 cm. They reported that UC and distribution efficiency of micro sprinkler were 60.8 and 36.4%, respectively. Sharma and Battawar[53] evaluated four commonly used UC methods: Christiansen method (CUC), Wilcox and Swailes (Ti), USDA pattern efficiency (PEU), and Barmi and Hore new coefficient (A). Higher values of UCs were observed in case of overlapping patterns along with higher crop yields. Average values of UC and PEU were close to each other. However, Christiansen method[4] gave a higher value of 64.8%. The increase in UC due to overlapping was 23.7% for UC and 100% for A. Values of UC and A under the test were close to those by Barmi and Hore.

Sakore[11] studied UC and wetted throw for three types of micro sprinklers. He found a UC of 60% for different spacings of micro sprinklers. Further, he observed an increase in UC spacing from 200 to 100 cm. Firake et al.[54] studied the effect of spacing and operating rate on UC. They reported the maximum UC (94.5%) at 1.62 m spacing of micro sprinkler with 33 lph discharge rate, whereas it was maximum (90%) at 1.8 m spacing for 57 lph discharge rate. They also found that EU of micro sprinkler for discharge rates of 33–57 lph ranged from 96 to 97.5%.

Buzescu et al.[55] studied "reconsideration of micro-sprinkler irrigation for vegetable crops and tested the MA-1 micro sprinkler for its suitability for simple irrigation or for irrigation plus plant protection, in the changed land ownership conditions in Romania". The MA-1 had a nozzle diameter of 2 mm, pressure of 1.5 bar, and discharge rate of 9.2 mm/h. It was tested for fixed position at a height of 2.3 m or mobile at a height of 0.8 m. The quality (uniformity) of distribution and its efficiency were compared for micro jet units at 70 and 120 lph. They concluded that micro sprinkling was an effective method, and it is suitable for use by small growers in Romania. Hills and Barragan[56] compared water application for drop tube sprayers, boom sprayers, and rotators, and they reported that the CUC was highest for the rotator emitter (94.60%) followed by drop tube sprayers (93.70%) and boom sprayers (89.50%). Wind speed up to 6.2 m/s had little effect on the UCs for the three types of sprinklers tested. The increases in the mean droplet diameters measured in the field for the overlapping patterns were as follows: drop tube sprayers (1.4 mm), boom sprayers (1.7 mm), and rotators (2.8 mm). It was concluded that the larger droplet diameters for the rotators might have been beneficial in maintaining uniform precipitation patterns under windy conditions. However, the larger droplets may lead to soil crusting on certain soils.

Tarjuelo et al.[57] determined the spatial distribution of water applied with solid set sprinkler systems in open field conditions. Factors affecting water distribution were direction and speed of wind, working pressure, design and number of nozzles, evaporation and drift losses, height of sprinkler above the ground, etc. Results showed that wind speed has a clear negative effect on irrigation uniformity. The larger the spacing between sprinklers, lower was irrigation uniformity. Uniformity increased by using jet-straightening vanes in the main nozzle, double nozzles under low wind speeds, and single nozzles under high wind speeds, and by locating the sprinkler at 2 m aboveground.

Schneider[58] found that application efficiencies for LEPA ranged from 95 to 98%. Spray application efficiencies exceeded 90% when runoff and deep percolation are negligible. Because of the start and stop nature of mechanical move irrigation systems, UCs for LEPA and spray are measured both along the irrigation system mainline and in the direction of travel. Along the mainline, UCs are generally in the 0.94–0.97 range for LEPA and in the 0.75–0.85 range for spray. In the direction of travel, the UCs are generally in the 0.75–0.85 range for LEPA with furrow diking and in the 0.75–0.90 range for spray. Rocha et al.[59] evaluated the performance of regulated and unregulated yellow models and unregulated violet model of mini sprinklers. Water DU, manufacturing variation coefficient, mean effective wetted radius, and water distribution profile were determined. The best sprinkler spacing for the vegetable crop was simulated. The analysis demonstrated that all models produced a manufacturing variation coefficient lower than 5%, classifying them in category A. On the basis of simulations, it was concluded that the emitter can work under the pressures of 250, 250, and 300 kPa, respectively, for the unregulated violet and regulated and unregulated yellow models.

Charles[60] reported that the Cal Poly ITRC irrigation evaluation programs have been widely used to assess the global DU of drip and micro sprayer irrigation systems. The system DU is estimated mathematically by combining the component DU values. The DU components include pressure differences, other causes (such as manufacturing variation, plugging, and wear), unequal drainage, and unequal application rates. Results were also presented from evaluations by several entities, including Cal Poly ITRC. Cal Poly evaluations of 329 fields provided an average

DU_{lq} of 0.80 for micro spray. Approximately 45% of the nonuniformity was due to pressure differences, 52% was due to other causes, 1.0% was due to unequal drainage, and 2% was due to unequal application rates. The data showed that it was possible to have high system DU values for at least a 20-year system life and also good design and management.

From the tests on residential irrigation systems, Melissa et al.[61] calculated the average low quarter distribution uniformity (DU_{lq}) value as 0.45. Rotary sprinkler resulted in significantly higher DU_{lq} compared with fixed pattern spray heads with a value of 0.49 and 0.41. Rotor heads had higher uniform distribution of 0.55 compared with 0.49 for spray head. Spray heads had better uniformity when fixed quarter circle nozzles were used as opposed to adjustable nozzles. Residential irrigation system uniformity can be improved by minimizing the occurrence of low pressure in the irrigation system and by ensuring that proper spacing is used in design and installation.

Barragan et al.[62] reported that EU has been one of the most frequently used criteria for micro irrigation design and evaluation. The original EU formula, which was derived by a worst combination of hydraulic variation and manufacturer's variation, can provide a very conservative design with a smaller value for EU than that measured in the field. They developed a revised formula for EU based on a statistical approach, and it provided more realistic value in the field. The revised formula increased the calculated values by 5–8% greater than the original values for EU. The values for EU can be applied to the hydraulic design using the same procedure as that proposed by the original EU.

Bansod and Shukla[63] conducted studies on five different types of micro sprinklers (coded for identification as MS-I, MS-II, MS-III, MS-IV, and MS-V) at three operating pressures (1.0, 1.5, and 2.0 kg/cm²) and at 35 cm riser height for obtaining the depth distribution pattern from single leg test. The depth distribution data was analyzed for CUC, Wilcox–Swailes uniformity coefficient (CUH), and Merriam and Keller DU. The decreased values of UCC, CUH, and DU were observed with increase in the spacing of micro sprinklers. At the rated pressure (2 kg/cm²), the maximum value of UCC (94.74%) was recorded for MS-V at 3×3 m micro sprinkler spacing. More than desired value, 70% of UCC was recorded only for MS-1 for all pressures under consideration and for spacing 3×3 m to 7×7 m, which indicated its superiority over other types of micro sprinklers. Overall performance of CUH was recorded for MS-V at 2.0 kg/cm² pressure for all micro sprinkler spacing. The highest CUH value (94.56%) was recorded for MS-III. A fairly good value of DU (78%) was recorded at all pressures and at 3×3 m and 4×4 m spacing for all the micro sprinklers except MS-IV.

Kishore et al.[64] concluded that the pressurized irrigation methods suffered from nonuniformity of water distribution. Unequal distribution is unacceptable for precision irrigation. It is now possible to combine small flow regulators that convert each sprayer into a pressure-compensated outlet (PCMS). The DU was 97.35%, and the

UC was 98.33%. An average yield of 6700 kg/ha was obtained with the PCMS irrigation compared with 6290 kg/ha under the sprinkler-irrigated area and 6300 kg/ha under the flood-irrigated area. The average weight of 1000 grains was 58 g for PCMS, 54.8 g from sprinkler, and 55.6 g from flood irrigation (control).

Jadhav et al.[65] found that a more precision was observed during manufacturing process for MSBK- and MSW-type micro sprinklers out of five micro sprinklers (MSBK, MSW, MSG, MSBL, and MSY). The EU and UC values for all the micro sprinklers ranged from 85.02 to 90.32% and 86.55 to 92.32%, respectively. For rotary sprinkler irrigation system, Jitander[22] determined UC, DU, and DC as per procedure given by Christiansen,[4] catch cans being at 2 m interval from sprinkler. The UC, DU, and DC were increased with nozzle size of sprinkler and pressure and were decreased with an increase in nozzle elevation.

5.3 MATERIALS AND METHODS

The present study was conducted at the instructional farm of College of Technology and Engineering to evaluate micro sprinkler operating characteristics under various operating conditions. During the study, the Christiansen method (catch can) was used to determine the UC,[4] DU, and DC. Keller and Karmeli[6] methods were used to determine precipitation patterns and EU. To develop the pressure discharge relationship for the micro sprinkler, nozzle discharge was collected in a container by turning micro sprinkler upside down. The experimental site was located at Udaipur, India (24°25′N latitude, 73°42′E longitude and at an elevation of 582 m above mean sea level). It is situated in Aravalli ranges of the southern part of Rajasthan. The area is characterized by subtropical continental semi-humid monsoon-type climate. The soil type of study area is sandy loam. The details about soil are given in Table 5.1.

TABLE 5.1 Soil Physical Properties

Property	Value
Soil type	Sandy loam
Percentage of proportion	Sand = 62.26%
	Silt = 19.04%
	Clay = 18.70%
Bulk density	1.52 gm/cc
Hydraulic conductivity	129.8 cm/day
Basic infiltration rate	2.3 cm/h
Field capacity	20.12%

The experimental setup consisted of centrifugal pump, water filter station (i.e., hydrocyclone filter, sand filter, and screen filter), fertilizer tank with various accessories (pressure gauge, non-return valve, air release valve, reducing elbow, nipple, bypass system), main line, sub-main line, lateral line, micro sprinklers with stake, control valve, flushing valve, and end cap. The experimental setup is shown in Fig. 5.1.

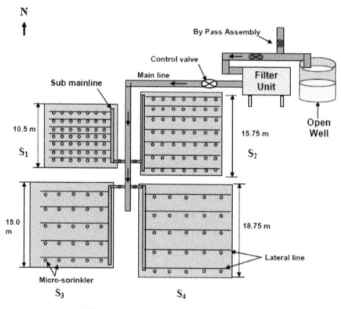

* Dimensions not to the scale.

Plot Notation	Plot size:	Micro-sprinkler spacing:
S_1	10.5 m x 10.5 m	1.5 m x 1.5 m
S_2	15.75 m x 15.75 m	2.25 m x 2.25 m
S_3	15.0 m x 15.0 m	3.0 m x 3.0 m
S_4	18.75 m x 18.75 m	3.75 m x 3.75m

FIGURE 5.1   Layout of the experimental setup.

For conducting experiment, an open well was used as the source of water. A 5 hp submersible pump was used to develop the required pressure for the experiment. The pump consisted of a bypass arrangement with flow control valve for controlling the operating pressure and corresponding flow rate in the lateral. A fertilizer tank of 90-L capacity was attached with the system, but it was not used for fertigation purpose. Filters were used to safeguard the system from dirty water and to protect sprinkler heads from being clogged. The type and the size of the filter depend upon the kind of dirt, diameter of sprinkler nozzles, its hourly discharge, and total amount of water per shift or cycle. For this purpose, hydrocyclone filter, sand filter, and screen filter were used in series, as shown in Fig. 5.2.

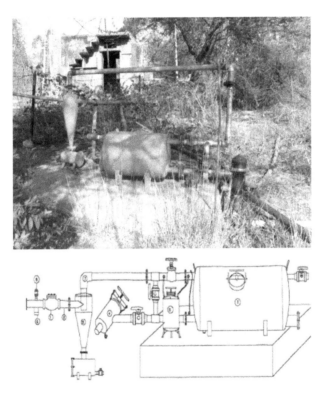

Sr. No.	ITEM
1	NON RETURN VALVE
2	HYDROCYCLONE FILTER
3	SAND FILTER
4	SCREEN FILTER
5	FERTILIZER TANK
6	AIR RELEASE VALVE
7	G.I. ELBOW/ REDUCING ELBOW
8	G.I. NIPPLE6' LONG
9	BY PASS ASSEMBLY

FIGURE 5.2 Filtration unit.

"Hydrocyclone filter" is also known as centrifugal filter or sand separator. It was used to separate the sand, silt, or particles heavier than water, but it cannot remove algae, fibers, clay, etc., present in water source. "Sand filter" was used to remove all types of physical impurities organic or inorganic, algae, silt, clay, suspended particles, etc. "Screen filter" is mandatory for every sprinkler irrigation system. Screen filter was used primarily for removing inorganic particles. A screen filter does not have the capacity to remove large amounts of suspended particles and organic particles without reducing the flow through the filter. Therefore, we need frequent flushing of a screen filter. Selection guide for filtration system is shown in Table 5.2.

TABLE 5.2   Detailed Specifications for Different Types of Filters

Filter Type	Nominal Flow Rate, m^3/h	Inlet–Outlet, inch	Maximum Operating Pressure, kg/cm^2	Gross Weight, kg	Use
Hydrocyclone filter	50	3"	10	54	Removal of sand and silt particles from water
Sand filter	50	3"	10	142	Removal of biological, organic, and physical impurities from water
Screen filter	25	2" Screen filter having 100 μm mesh size	10	3.70	Removal of physical contaminants from water

The PVC pipe was used as main line of size 75 mm (OD) and sub-main line of size 63 mm (OD) for micro sprinkler irrigation system. The low-density polyethylene black color as lateral line of size 16 mm (OD) was used. This is the most commonly used size of laterals for micro irrigation. Micro tube of 4 mm (OD) was used to connect micro sprinkler with lateral line (Fig. 5.5). In addition to main components of the sprinkler system, various fittings and accessories were used as an essential part of the system (Figs. 5.3 and 5.4). The accessories were gate valve, pressure gauge, pitot tube, non-return valve, air release valve, reducing elbow, and nipple, bypass, and end cap, etc.

FIGURE 5.3 Control valve (Left).

FIGURE 5.4 Pressure gauge.

FIGURE 5.5   Micro sprinkler along with lateral connected by micro tube.

Micro sprinkler nozzle (Fig. 5.5) is the most important component of micro sprinkler irrigation system. Its operating characteristics under optimum water pressure and climate conditions determine its suitability and efficiency. Specifications of micro sprinkler for this study are as under.

Make : Jain irrigation system Ltd.
Stake height : 25 cm
Nozzle size : 1.12 mm
Other details

Operating pressure, kg/cm^2	Wetting radius, m	Discharge, lph
1.0	4.6	44
1.5	5.1	55
2.0	6.9	63
2.5	7.2	71

5.3.1 EXPERIMENTAL LAYOUT

The pressure–discharge relationship, UC, EU, and DU were evaluated at operating pressures ranging from 0.75 to 2.0 kg/cm^2 (with an increment of 0.25 kg/cm^2). Further for precipitation patterns were also studied at operating pressure range of 0.75–2.5 kg/cm^2, for the micro sprinkler spacing of 1.5 × 1.5 m, 2.25 × 2.25 m, 3.0 × 3.0 m, and 3.75 × 3.75 m, and under low-wind-speed conditions ranging from 0 to 6.5 km/h.[66] The manufacturer recommended operating pressures between 1.0 and 2.5 kg/cm^2. All tests were carried out during 7:00–9:00 a.m. and 5:00–7:00 p.m. to reduce evaporation losses due to temperature changes.

The experimental area was divided into four sectors (S$_1$, S$_2$, S$_3$, and S$_4$), each having a plot size of 10.5 × 10.5 m, 15.75 × 15.75 m, 15.0 × 15.0 m, and 18.75 × 18.75 m, respectively. In plots S$_1$, S$_2$, S$_3$, and S$_4$, micro sprinkler spacings were 1.5 × 1.5 m, 2.25 × 2.25 m, 3.0 × 3.0 m, and 3.75 × 3.75 m, respectively (Figs. 5.1 and 5.6). For the lower spacings (i.e., 1.5 × 1.5 m and 2.25 × 2.25 m), each plot was provided with seven laterals, and each lateral had seven micro sprinklers. Similarly, for higher spacing (i.e., 3.0 × 3.0 m and 3.75 × 3.75 m), each plot was provided with five laterals, and each lateral was equipped with five micro sprinklers. Each spacing was replicated three times (R$_1$, R$_2$, and R$_3$).

To minimize the effect of wind, three sides of study area was covered with polyethylene sheets as a wind breaker of 2 m in height and fourth side was already surrounded by large vegetative area (Fig. 5.6).

FIGURE 5.6 Positions of catch cans at grid points.

5.3.2 GRID SELECTION AND PLACEMENT OF CATCH CANS

Walker[67] suggested to place catch cans at 0.5 m interval at two or four radials for each micro sprinkler. He assumed that the application rate for a given distance from the head is uniform around the head. However, this assumption may not be true in practice. Post et al.[50] recommended placement of catch cans in a square matrix pattern around the sprinkler head. In the present study, the recommendation of Post et al.[50] was followed. For this purpose, grid of 50 cm × 50 cm was formed, and the precipitation was collected in catch cans placed at grid points. These observations were used to evaluate UC (%), DC (%), and depth of applications (d, mm/h). For each plot, a centrally located representative area was demarked, where precipitation of all adjacent micro sprinklers was received.

5.3.3 PRESSURE–DISCHARGE RELATIONSHIPS

The ideal micro sprinkler irrigation system is one in which all the sprinkler heads deliver the same volume of water in a given irrigation duration. From the practical point of view, it is impossible to achieve this idealized performance requirement, because the nozzle flow is affected by variation in water pressures in the sub-mains and lateral lines.

The flow variations caused by water pressure variation in a micro sprinkler system is termed as hydraulic variation. The hydraulic characteristics of a spray nozzle significantly affect the following aspects of irrigation performance: the wetting pattern and water distribution, and variation in nozzle discharge at varying operating pressures. In addition to this, the water network flows and operating pressures are directly determined by the nozzle pressure and discharge relationship. Therefore, designer and user must have a clear understanding of nozzle characteristics. In general, the flow rate and throw of nozzle are affected by hydraulic pressure at the nozzle and flow path dimensions of the sprinkler head. The following pressure–discharge relationship for a sprinkler has been proposed by Keller and Karmeli[6]:

$$q = KP^x \tag{6}$$

where q is the flow rate; K is the characteristic discharge coefficient of the nozzle, which also depends on the choice of units for q and P; P is the operating pressure; and x is the nozzle exponent characterizing the nozzle flow regime.

The values of K and x in eq 6 can be determined by fitting a logarithmic equation to the data from the field or manufacturer. With a suitable curve-fitting computer program, a curve can be fitted to data points, and pressure versus discharge graphs can be plotted. Alternatively the values of "x" and "K" can be determined analytically by using the following equation by Keller and Karmeli[6]:

$$\text{Log } q = \text{Log } K + x \text{ Log } P \text{ or } Y = A + m\,C \tag{7a}$$

$$X = \frac{\log(q_1 / q_2)}{\log(P_1 / P_2)} \tag{7}$$

where q_1 and q_2 are discharges at pressures P_1 and P_2, respectively. This value of "x" can be used to solve for K. Reader may note that eq 7a is a straight line. In the present analysis, in addition to the analytical determination of K and x, the best fit curve between pressure and discharge was obtained by using nonlinear regression technique. For obtaining pressure and discharge relationships, each of the micro sprinkler was operated for specified duration of 2.0 min, and discharge through nozzle was collected by keeping the micro sprinkler upside down in a plastic container (Fig. 5.7). Volume of water was measured with a graduated cylinder, and then it was converted into liter per hour. Operating pressures of 0.75, 1.0, 1.25, 1.50, 1.75, and 2.0 kg/cm² were selected. The desired operating pressure was maintained with the help of bypass arrangement. Each test was repeated three times to have a representative average value.

FIGURE 5.7   Collection of a water sample from the micro sprinkler at a given pressure.

5.3.4 PRECIPITATION PATTERN

Catch cans were placed at the grid of 0.5 × 0.5 m, formed around the micro sprinklers to determine the precipitation pattern (Fig. 5.6). First cans were placed at half spacing in all four principal directions. Micro sprinkler was operated for half an hour for each test pressure. Volume of water collected in catch can was converted into depth of precipitation per unit time (mm/h for all pressures under consideration). The precipitation pattern was studied by drawing the contour of equal precipitation depth (isohyetal lines) and distribution profile.

5.3.4.1 PRECIPITATION CHARACTERISTICS OF THE MICRO SPRINKLER

Precipitation characteristics of the micro sprinkler were studied in terms of precipitation pattern, distribution profile, and precipitation performance parameters such as effective radius, m; average application depth, mm/h; effective maximum depth, mm/h; absolute maximum depth, mm/h; mean application depth, mm/h; precipitation DC, %; and CV.

5.3.4.1.1 EFFECTIVE RADIUS

Effective radius is the distance between the center of a micro sprinkler and the point at which the profile meets the horizontal axis on precipitation pattern.[30]

5.3.4.1.2 AVERAGE APPLICATION DEPTH (MM/H)

Keller and Merriam[30] used concentric ring approach to calculate average application depth. The total wetted area was divided into four concentric rings. The radius of each ring was taken as 0.4, 0.6, 0.78, and 0.93 times the wetted radius, respectively, from inner to outer periphery. The radii were used to draw the concentric rings. The average depth in each ring was first estimated, and the average precipitation depth was calculated as given below:

$$\text{Average application depth} = \frac{d_1 + d_2 + d_3 + d_4}{4} \tag{8}$$

where d_1, d_2, d_3, and d_4 are average depths in first, second, third, and fourth concentric rings, respectively, around the micro sprinkler.

5.3.4.1.3 EFFECTIVE MAXIMUM DEPTH (MM/H)

Effective maximum depth is the average value of 5% of total cans having maximum collection quantity and is expressed in mm/h.

5.3.4.1.4 ABSOLUTE MAXIMUM DEPTH (MM/H)

Absolute maximum depth is the maximum of all observations.

5.3.4.1.5 MEAN APPLICATION DEPTH (MM/H)

Mean application depth is the average depth between the areas constrained by effective radius.

5.3.4.2 DISTRIBUTION CHARACTERISTIC

This term characterizes the water distribution of single sprinkling device. It was calculated as follows:

$$\text{Distribution characteristic} = \frac{\text{Area receiving the depth greater than average depth}}{\text{Total wetted area}} \times 100 \qquad (9)$$

5.3.4.3 COEFFICIENT OF VARIATION

The CV was calculated by using the following equations by Keller and Merriam[30]

$$CV = \frac{\sigma}{da} \times 100 \qquad (10)$$

$$\sigma = \sqrt{\frac{\sum X^2}{N}} \qquad (11)$$

where da is the mean application depth; $\sum X^2$ is the deviation from mean depth, mm; and N is the number of observations.

5.3.5 UNIFORMITY OF WATER APPLICATION

The uniformity of water application of a micro sprinkler system is dependent on the DC of a sprinkler nozzle. There are procedures for evaluating application uniformity. Christiansen formula[4] is specific for overlapping sprinklers and is used to calculate UC. The procedure proposed by Keller et al.[30] for overlapping and slightly overlapping sprinklers can be used to calculate DU. High efficiency in the operation of an irrigation system is not necessarily economical. Therefore, irrigation system should be evaluated to rationally decide whether the system should be modified or a different system should be adopted. Efficiencies based on field data have an accuracy of ±5%. Therefore, the irrigation system must be evaluated under maximum possible ideal conditions in order to have information regarding the performance of the system.

5.3.5.1 UNIFORMITY COEFFICIENT

A measurable index of the degree of uniformity of micro sprinkler operating under given sets of conditions is known as the UC. The UC is affected by the pressure discharge relationship and sprinkler spacing. The UC was determined by the following equation by Christiansen[4]:

$$UC = UC = \left[1 - \frac{\sum X}{mn}\right] \times 100 \qquad (12)$$

where UC is the uniformity coefficient, %; m is the average value of all observations, mm, Fig. 5.8; n is the total number of observation points; and X is the numerical deviation of individual observation from the average application rate, mm.

FIGURE 5.8 Catch can to receive precipitation for studying performance parameters.

5.3.5.2 EMISSION UNIFORMITY

EU is the measure of the uniformity of discharge from all micro sprinklers and is the single most important parameter for evaluating performance. EU shows relationship between minimum and average values of discharge. EU is needed for calculating gross depth of irrigation, irrigation interval, and required system capacity. It depends upon water temperature and the manufacturer's CVs of the system. The following equation by Keller and Karmeli[6] was used to calculate EU:

$$EU = 100 \times \frac{q_n}{q_a} \qquad (13)$$

where EU is the emission uniformity, %; q_n is the average of the lowest quarter of the emission point discharges for field data, lph; and q_a is the average emission point discharges of test sample operated at the reference pressure head, lph.

5.3.5.3 DISTRIBUTION UNIFORMITY (DU)

The DU (%) indicates the degree to which the water is applied uniformly over a given area. DU was determined as follows:

$$DU = \frac{\text{Average low quarter depth of water caught}}{\text{Average depth of water caught}} \times 100 \qquad (14)$$

5.4   RESULTS AND DISCUSSION

5.4.1   PRESSURE–DISCHARGE RELATIONSHIP

The discharge rates of micro sprinklers (Fig. 5.8) were recorded at operating pressures ranging from 0.75 to 2.0 kg/cm^2 with an increment of 0.25 kg/cm^2. The average values obtained are reported in Table 5.3. The minimum discharge was 32.99 lph at 0.75 kg/cm^2, and the maximum discharge was 73.10 lph at 2.0 kg/cm^2 operating pressure (Table 5.3). The discharge was increased with increase in operating pressure. These findings are in close agreement with those reported by Firake and Salunkhe,[10] Gawali and Budhan,[15] and Suryawanshi et al.[17] The pressure–discharge relationships based on the following equation by Keller and Karmeli[6] are shown in Figs. 5.9–5.13, at different spacing of micro sprinklers.

$$Q = K \cdot P^x \qquad (15)$$

where Q is the discharge of a micro sprinkler, lph; P is the operating pressure, kg/cm^2; K is the characteristic constant; and x is the discharge exponent (Table 5.4).

TABLE 5.3   Average Discharge of a Micro Sprinkler at Corresponding Operating Pressure

Operating Pressure	Discharge			Average Discharge
	Test Run			
	I	II	III	
kg/cm^2			lph	
0.75	33.13	32.96	32.87	32.99
1.00	46.08	46.28	46.28	46.21
1.25	57.30	57.98	58.10	57.79
1.50	62.83	62.93	62.93	62.90
1.75	66.42	66.38	66.42	66.40
2.00	73.17	73.20	72.92	73.10

TABLE 5.4   The Value of (x) Characterizes the Flow Regime[6]

Value of "x"	Flow Regime
$x = 0.5$	Fully turbulent
$0.5 < x < 0.8$	Partially turbulent
$0.8 < x < 1.0$	Unstable flow regime
$x = 1.0$	Laminar flow

Characteristic constant for micro sprinkler was 44.45, whereas discharge exponent was 0.78. High value of R^2 (=0.95) indicates that the regression coefficients were significant and there is high goodness of fit. The relationships between dependent variable (discharge) and independent variable (operating pressure) are shown in Fig. 5.9. In designing pressurized irrigation systems, general accepted limit is 20% variation in operating head and corresponding 10% discharge variation. According to these criteria, the micro sprinkler performance under study can be considered as satisfactory as the value of discharge exponent observed was 0.79.

An attempt was also made to establish functional relationships between the operating pressure and discharge as influenced by the spacing of micro sprinklers (Figs. 5.10–5.13). For this purpose, average overlap of 50% of the micro sprinklers for the given spacings was considered. The relationships are shown in Figs. 5.10–5.13 for micro sprinkler spacing of 1.5 × 1.5 m, 2.25 × 2.25 m, 3.0 × 3.0 m, and 3.75 × 3.75 m, respectively. The values of characteristic constants, discharge exponents, and goodness of fit as influenced by micro sprinkler spacing are reported in Table 5.5.

The micro sprinkler characteristic constants were 44.91, 44.46, 44.43, and 43.99, and discharge exponents were 0.77, 0.78, 0.78, and 0.79 at corresponding spacing of 1.5×1.5 m, 2.25×2.25 m, 3.0×3.0 m, and 3.75×3.75 m, respectively. The higher value of R^2 greater than 0.95 indicated the goodness of fit.

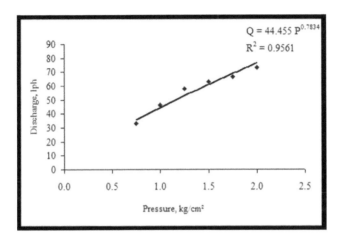

FIGURE 5.9   Pressure–discharge relationship for a micro sprinkler.

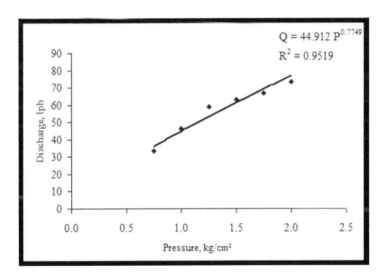

FIGURE 5.10   Pressure–discharge relationship for 1.5×1.5 m micro sprinkler spacing.

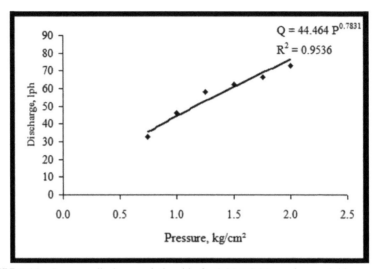

FIGURE 5.11   Pressure discharge relationship for 2.25 × 2.25 m micro sprinkler spacing.

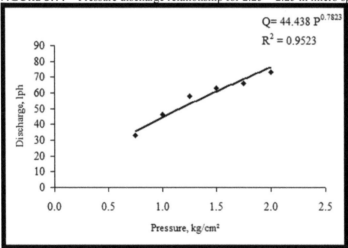

FIGURE 5.12   Pressure–discharge relationship for 3 × 3 m micro sprinkler spacing.

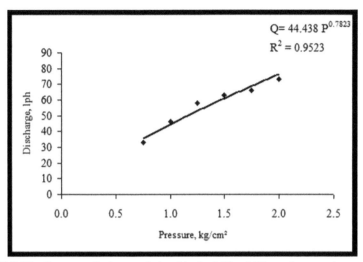

FIGURE 5.13   Pressure–discharge relationship for 3.75×3.75 m micro spacing.

5.4.2   PRECIPITATION PATTERN

The objective of irrigation is to have uniform distribution of water. Under sprinkler method of irrigation, spray pattern depends on precipitation distribution, sprinkler position in the field, type of sprinkler head, operating pressure, field topography, etc. These variables influence the overall uniform distribution of irrigation water. The efforts are always made to have the optimum operating pressure to get the maximum uniform distribution resulting in maximum efficiencies. To achieve the same, sprinklers were operated during the calm hours of the day (i.e., 7–9 a.m. and 5–7 p.m.) to avoid spray and drift losses. The precipitation distribution pattern can be studied in terms of studying its distribution profile as a function of operating pressure as well as performance parameters such as effective radius, m; average application depth, mm/h; effective maximum depth, mm/h; absolute maximum depth, mm/h; mean application depth, mm/h; and CV, %. In this study, the micro sprinkler was tested for operating pressure in the range of 1.0–2.5 kg/cm² with an increment of 0.5 kg/cm².

5.4.2.1   DISTRIBUTION PROFILE

For the determination of precipitation distribution patterns, contours of equal precipitation were drawn at an interval of 1.5 mm. The patterns along with corresponding distribution profiles are presented at operating pressures of 1.0 (Figs. 5.14 and 5.15), 1.5 (Figs. 5.16 and 5.17), 2.0 (Figs. 5.18 and 5.19), and 2.5 (Figs. 5.20 and 5.21) kg/cm². From the precipitation contours, it was observed that precipitation was concentrated nearer to the micro sprinkler because of closer spacing between the contours near to the center and wider toward the periphery. It was further noticed that contours are more or less equi-spaced with increase in operating pressure, which can be seen from the comparison of precipitation contour patterns for 1.0 and 2.5 kg/cm² operating pressures shown in Figs. 5.14 and 5.20, respectively.

The pressure distribution was also studied in terms of precipitation profiles. The distribution profiles are shown in Figs. 5.15, 5.17, 5.19, and 5.21 at operating pressures of 1.0, 1.5, 2.0, and 2.5 kg/cm², respectively. At a lower operating pressure (1.0 kg/cm²), the distribution profile indicated flat elliptical shape (Fig. 5.15), which gradually changed to triangular profile. This type of profile requires closer spacing for more overlapping, uniform average depth, and higher UC. It can be concluded that the closer spacing of micro sprinklers should be adopted for an operating pressure of 1.0 kg/cm² under low wind speed to get higher uniformity of water distribution. However, the cost of sprinkler system will increase.

The precipitation patterns in Figs. 5.15, 5.17, 5.19, and 5.21 indicate that the increase in operating pressure was able to increase the flatness in the precipitation curve up to 2.0 kg/cm². However, at an operating pressure of 2.5 kg/cm², precipitation curve significantly changed its shape from elliptical to triangle. This indicates that if operating pressure is increased beyond 2.0 kg/cm², the centroid contractual precipitation pattern is obtained, which may reduce the uniformity of water application. To reduce the initial cost of micro sprinkler system, the operating pressure ranging from 1.5 to 2.0 kg/cm² was found to be suitable to enhance the spacing between micro sprinklers.

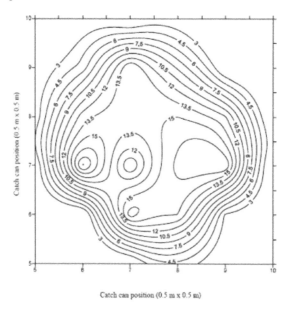

Catch can position (0.5 m x 0.5 m)

FIGURE 5.14 Precipitation pattern of micro sprinkler at an operating pressure of 1.0 kg/cm².

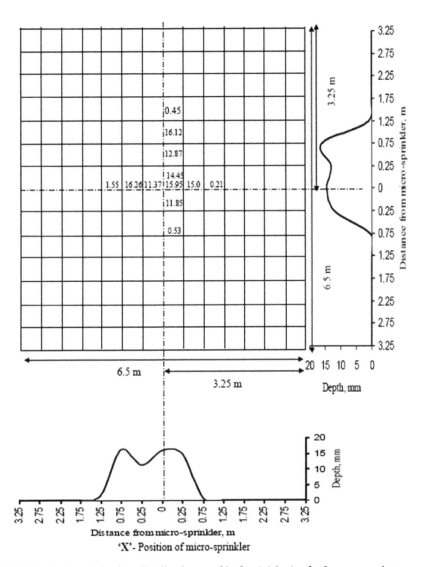

FIGURE 5.15   Precipitation distribution profile for 1.0 kg/cm² of pressure along XX- and YY-axes.

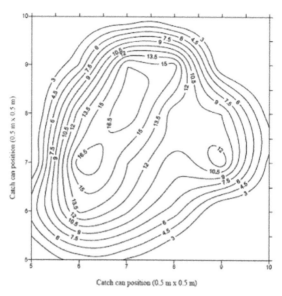

FIGURE 5.16   Precipitation pattern of micro sprinkler at an operating pressure 1.5 kg/cm².

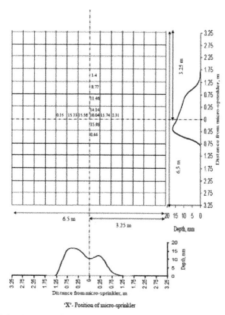

FIGURE 5.17   Precipitation distribution profile for 1.5 kg/cm² of pressure along XX- and YY-axes.

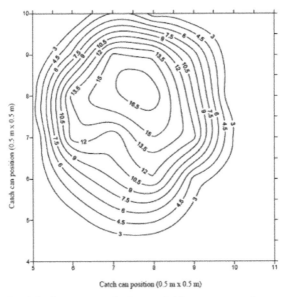

FIGURE 5.18   Precipitation pattern of micro sprinkler at an operating pressure of 2.0 kg/cm².

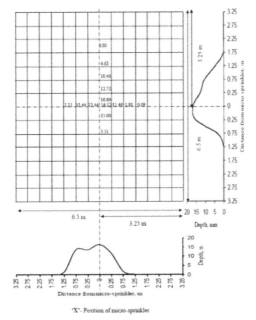

FIGURE 5.19   Precipitation distribution profile for 2.0 kg/cm² of pressure along XX- and YY-axes.

FIGURE 5.20 Precipitation pattern of micro sprinkler at an operating pressure of 2.5 kg/cm².

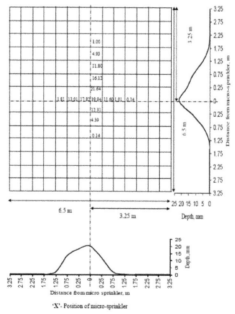

FIGURE 5.21 Precipitation distribution profile for 2.5 kg/cm² of pressure along XX- and YY-axes.

5.4.2.2   PRECIPITATION PERFORMANCE PARAMETERS

Keller and Merriam[30] have given certain parameters to study the precipitation patterns. The sample calculations for parameters for precipitation performance are presented in Appendix I. The results of precipitation performance are reported in Table 5.7 for different operating pressures.

TABLE 5.5   Micro Sprinkler Characteristic Constants (K), Discharge Exponents (x), and Goodness of Fit (R^2) for Different Spacings of Micro Sprinkler

Spacing, m	K	x	R^2
1.5 × 1.5	44.912	0.7749	0.9519
2.25 × 2.25	44.464	0.7831	0.9536
3.0 × 3.0	44.438	0.7823	0.9523
3.75 × 3.75	43.996	0.7936	0.9646

TABLE 5.6   Precipitation Characteristics as Influenced by Operating Pressure Based on Keller Method

Precipitation Characteristics	Operating Pressure, kg/cm^2			
	1.0	1.5	2.0	2.5
Absolute maximum depth, mm/h	19.80	17.54	17.26	21.78
Average precipitation depth, mm/h	7.69	7.73	8.46	9.23
Coefficient of variation, %	0.67	0.77	0.86	0.83
Distribution characteristic, %	24.07	17.20	18.70	36.27
Effective maximum depth, mm/h	15.56	14.14	14.00	14.95
Effective radius, m	1.0	1.25	1.5	1.75
Mean application depth, mm/h	11.40	8.08	6.72	7.62

TABLE 5.7 Uniformity Coefficient (UC) at Different Operating Pressures and Micro Sprinkler Spacings

Operating Pressure (kg/cm²)	Spacing (m)	Uniformity Coefficient, UC (%)			Average UC (%)
		I	II	III	
0.75	1.5 × 1.5	79.62	78.85	83.10	80.52
	2.25 × 2.25	71.76	69.3	69.42	70.16
	3.0 × 3.0	62.58	67.21	70.43	66.74
	3.75 × 3.75	31.26	35.19	37.17	34.54
1.00	1.5 × 1.5	83.63	80.61	81.63	81.96
	2.25 × 2.25	71.96	73.52	73.33	72.94
	3.0 × 3.0	73.32	66.04	71.49	70.28
	3.75 × 3.75	44.16	40.07	43.19	42.47
1.25	1.5 × 1.5	80.21	85.35	83.40	82.99
	2.25 × 2.25	78.28	75.09	79.55	77.64
	3.0 × 3.0	75.62	73.04	75.40	74.69
	3.75 × 3.75	44.51	48.45	46.59	46.52
1.50	1.5 × 1.5	85.06	84.23	84.10	84.46
	2.25 × 2.25	86.31	80.85	77.34	81.50
	3.0 × 3.0	75.29	78.59	76.46	76.78
	3.75 × 3.75	42.06	52.53	59.49	51.36

	1.5 × 1.5	87.20	84.88	84.52	85.53
	2.25 × 2.25	82.63	81.57	81.59	81.93
1.75	3.0 × 3.0	78.93	76.23	77.75	77.64
	3.75 × 3.75	55.06	60.13	59.34	58.18
	1.5 × 1.5	85.94	85.77	85.60	85.77
	2.25 × 2.25	86.50	80.31	85.12	83.98
2.00	3.0 × 3.0	78.72	79.30	78.13	78.72
	3.75 × 3.75	61.62	63.28	63.13	62.68

5.4.2.2.1   EFFECTIVE RADIUS (RE)

Effective radius ranged from 1.0 to 1.75 m at 1.0–2.5 kg/cm² operating pressure. In Keller method, distribution profile is required to extend until it cuts horizontal axis and estimates comparatively lower effective radius. Findings of the present study are in agreement with those reported by Pandey et al.,[29] Pampattiwar et al.,[19] and Patil et al.[68] Thus, under low-wind-speed conditions and for 50% overlapping, the spacing between micro sprinkler at 1.0 kg/cm² should be 1.0 m and compared with 1.75 m at a pressure of 2.5 kg/cm². However, the system operation at low pressure (1.0 kg/cm²) increases cost, and high pressure (2.5 kg/cm²) may reduce uniformity of precipitation pattern. Thus, if the system operates between an operating pressure of 1.5 and 2.0 kg/cm², the maximum spacing can run from 1.25 to 1.5 m with 50% overlapping under low-wind-speed condition.

5.4.2.2.2   AVERAGE APPLICATION DEPTH

The maximum average application depth of water was 9.23 mm/h at 2.5 kg/cm² of operating pressure. At 1.0, 1.5, and 2.0 kg/cm² of operating pressure, the average application depths were 7.69, 7.73, and 8.46 mm/h, respectively.

5.4.2.2.3   EFFECTIVE MAXIMUM DEPTH

Five percent of total cans receiving maximum precipitation were considered for the estimation of effective maximum depth. The estimated values were 15.56, 14.14, 14.00, and 14.95 mm/h at 1.0, 1.5, 2.0, and 2.5 kg/cm² of operating pressures, respectively (Table 5.6). Highest effective maximum depth was 15.56 mm/h at 1.0 kg/cm² operating pressure.

5.4.2.2.4   ABSOLUTE MAXIMUM DEPTH

Absolute maximum depth is the maximum depth that is recorded among all observation cans.[30] Observed values are reported in Table 5.6. These values were 19.80, 17.54, 17.26, and 21.78 mm/h at 1.0, 1.5, 2.0, and 2.5 kg/cm² operating pressures, respectively. Maximum depth was found at 2.5 kg/cm² operating pressure.

5.4.2.2.5   MEAN APPLICATION DEPTH

Mean application depth is the average depth recorded between the area constrained by the effective radius and estimated by the procedure in this study. Mean depth were 11.40, 8.08, 6.72, and 7.62 mm/h at operating pressures of 1.0, 1.5, 2.0, and 2.5 kg/cm², respectively.

5.4.2.3   DISTRIBUTION CHARACTERISTIC

Distribution characteristic is a ratio of wetted area receiving the depth greater than average depth to the total wetted area and is expressed as percentage.[30] The values estimated from the observed data are reported in Table 5.6. These values are 24.07, 17.20, 18.70, and 36.27%, respectively. Further micro sprinkler at 1.5 kg/cm² and 2.0 kg/cm² pressure yielded values of 17.2 and 18.7%, indicating that the wetted area received depth that was greater than the average depth. Findings confirm that the distribution profile is elliptical at lower operating pressure, and it gradually changed to triangular type with increase in operating pressure. The findings also indicated that micro sprinkler under study should be operated within the range of 1.5–2.0 kg/cm² of pressure.

5.4.2.4   COEFFICIENT OF VARIATION

The CV describes the deviation of observed values of depth from the mean depth in all the cans that were located within the effective radius.[6] The values estimated for different operating pressures are reported in Table 5.6. Minimum value of CV was 0.67 at 1.0 kg/cm² of operating pressure. When the pressure was increased to 2.0 and 2.5 kg/cm², the estimated values were 0.86 and 0.83, respectively. Similarly, for a low operating pressure of 1.5 kg/cm², the estimated value was 0.77. Thus, these results further confirm the superiority of the precipitation pattern of micro sprinkler under study for the operating pressure of 1.0 kg/cm² compared with other operating values.

The precipitation characteristics in terms of effective maximum depth and absolute maximum depth indicate more deviation of absolute maximum depth from effective maximum depth if the system is operated at a pressure of 2.5 kg/cm². This indicates more variations in precipitation at this pressure. This result can be attributed to the results obtained for distribution profile.

The above discussion on precipitation characteristics as influenced by operating pressure indicates that the micro sprinkler system at a pressure between 1.5 and 2.0 kg/cm² is more suitable for the spacing range of 1.25–1.5 m to achieve 50% overlapping. With the spacing of 1.5 × 1.5 m operating at 2.0 kg/cm², the number of sprinklers per hectare can be reduced by 55% compared with the number of sprinklers required for the spacing of 1.0 × 1.0 m operating at 1.0 kg/cm². If the system is operated at 1.5 kg/cm², the spacing required is 1.25 × 1.25 m, which reduces the number of sprinkler by 36% compared with the number of sprinklers required for the spacing of 1.0 × 1.0 m. Therefore, the cost of laterals can be saved through the saving of lateral length per hectare by 33% and 20% by using the spacing of 1.5 × 1.5 m and 1.25 × 1.25 m, respectively, compared with the spacing of 1.0 × 1.0 m.

5.4.3   UNIFORMITY COEFFICIENT

Precipitation collected in catch cans placed at a grid point from the representative area is measured to estimate the UC as shown in Table 5.7. Appendix II shows an example of observation of precipitation depths (mm) for 30 min at grid points to calculate UC and DU for 1.5 × 1.5 m spacing at 0.75 kg/cm².

Figure 5.22 shows effects of operating pressures and different micro sprinkler spacings on UCs (%). It can be observed that the UC was 80.52% at an operating pressure of 0.75 kg/cm², for micro sprinkler spacing of 1.5 × 1.5 m. At a spacing of 2.25 × 2.25 m, the UC was 70.16%. Further increase in spacing (3.0 × 3.0 m) lowered the value of UC to 66.74%. The reduction was 3.42%. However, with further increase in spacing to 3.75 × 3.75 m, the UC was 34.54% showing a drastic reduction of 32.20%.

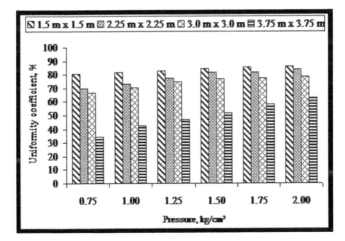

FIGURE 5.22 Uniformity coefficients (UCs, %) at different operating pressures for different spacings of micro sprinklers.

For an operating pressure of 1.0 kg/cm², the UC was 81.96% for a micro sprinkler spacing of 1.5 × 1.5 m compared with 72.94% at a spacing of 2.25 × 2.25 m. Further increasing the spacing to 3.0 × 3.0 m and 3.75 × 3.75 m, the UC was 70.28 and 42.47%, and the reductions were 2.66 and 27.81%, respectively. For an operating pressure of 1.5 kg/cm², the reduction was 25.42% when the spacing was changed from 3.0 × 3.0 m to 3.75 × 3.75 m with the UC values of 76.28 and 51.46%, respectively. Similar trend of reduction in UC was noticed when operating pressures were further increased from 1.25 to 2.0 kg/cm² with an increment of 0.25 kg/cm².

It was further revealed that for a given spacing, the UC was affected with operating pressure. For 0.75 kg/cm² operating pressure and 1.5 × 1.5 m spacing, the estimated UC was 80.52%. When the pressure was changed to 1.0 kg/cm² with the micro sprinkler spacing of 1.5 × 1.5 m, the UC was improved to 81.96%. Similar trends were observed for other operating pressures for all spacings under study. Thus, the overall results indicated that UC was improved with increase in operating pressures for a given spacing.

To study the combined effect of operating pressures and spacings of micro sprinkler, the data was statistically analyzed. The analysis indicated the significant effect of operating pressure and spacings of micro sprinkler on UC as shown in Table 5.8.

TABLE 5.8 ANOVA Showing Effects of Operating Pressures and Spacings of Micro Sprinkler on Uniformity Coefficients (UCs, %)

Source	DF	SS	MS	F Value	SE (m)	CD (5%)	CD (1%)
A = operating pressure	5	1877.01	375.401	46.057[a]	0.824	2.343	3.127
B = micro sprinkler spacing	3	12374.5	4124.84	506.061[a]	0.6729	1.913	2.554
A × B	15	572.841	38.1894	4.685[a]	1.648	4.687	6.255
Error	48	391.242	8.15087				

[a]Significant at 1%.

The relationships between operating pressure and UC for different micro sprinkler spacings are shown in Fig. 5.22. The relationships were nonlinear and followed power law, as shown in Table 5.9 and below:

$$UC = m \, P^n \tag{16}$$

where UC is the uniformity coefficient; m is the characteristic constant; P is the operating pressure in kg/cm²; and n is the exponent.

Increasing micro sprinkler spacings increased the value of exponent from 0.03 to 0.32 and decreased the value of characteristic constant from 80.18 to 33.85. The correlation coefficient (R^2) was also increased from 0.96 to 0.98. The results of correlation of coefficient (R^2) indicated that the UC followed power law for all micro sprinkler spacings (Table 5.9). The values of exponent indicate the variation of UC with respect to operating pressure. The lower values indicate less variation of UC with operating pressure and vice versa. The characteristic constant was highest for the spacing of 1.5 × 1.5 m, indicating at least 80% of UC at 1.0 kg/cm² pressure, which is acceptable for desirable performance of the system. This confirms that the spacing of 1.5 × 1.5 m is suitable for micro sprinkler compared with higher spacing.

The results in Table 5.9 indicate that the suitable operating pressure was for the spacing of 1.5 × 1.5 m as pressure was changed from 1.5 to 2.0 kg/cm².

TABLE 5.9   Values of Micro Sprinkler Characteristic Constant (m), Exponent (n), and Goodness of Fit (R^2) for Different Spacings

Spacing, m	m	n	R^2
1.5 × 1.5	80.181	0.037	0.9686
2.25 × 2.25	69.32	0.1059	0.9646
3.0 × 3.0	66.56	0.0967	0.9829
3.75 × 3.75	33.85	0.3251	0.9808

5.4.4 EMISSION UNIFORMITY

The values of EU (%) of the micro sprinkler system at different operating pressures were calculated using the procedure as described by Keller and Karmeli[6] and are reported in Table 5.10 and Fig. 5.23. It was revealed that the average value of EU in all the treatments ranged from 96.13 to 94.96% at an operating pressure of 0.75 kg/cm². The values were gradually decreased when spacing was increased from 1.5 × 1.5 m to 3.75 × 3.75 m. The micro sprinkler discharge was taken into account for the calculation of EU. The EU was mainly influenced by operating pressure and not by the spacing of micro sprinkler. To achieve maximum EU, higher pressure is desirable.

TABLE 5.10 Estimated Values of Emission Uniformity (EU, %) at Different Operating Pressures for Micro Sprinkler Spacings

Operating Pressure (kg/cm²)	Spacing (m)	Emission Uniformity, EU (%)			Average EU (%)
		I	II	III	
0.75	1.5 × 1.5	97.40	95.60	95.39	96.13
	2.25 × 2.25	95.34	96.18	95.69	95.73
	3.0 × 3.0	93.82	95.28	96.65	95.25
	3.75 × 3.75	95.53	94.37	94.97	94.96
1.00	1.5 × 1.5	97.35	95.95	95.78	96.36
	2.25 × 2.25	95.72	95.31	97.68	96.24
	3.0 × 3.0	97.16	94.61	94.38	95.38
	3.75 × 3.75	95.53	94.18	95.77	95.16
1.25	1.5 × 1.5	96.83	96.06	96.93	96.61
	2.25 × 2.25	96.74	96.66	95.97	96.46
	3.0 × 3.0	96.37	95.79	94.18	95.45
	3.75 × 3.75	99.22	93.89	93.02	95.38
1.50	1.5 × 1.5	97.46	97.11	97.88	97.48
	2.25 × 2.25	96.79	97.13	96.49	96.80
	3.0 × 3.0	94.99	95.80	97.08	95.96
	3.75 × 3.75	94.11	93.75	98.67	95.51
1.75	1.5 × 1.5	97.47	98.06	97.71	97.75
	2.25 × 2.25	98.52	97.38	95.75	97.22
	3.0 × 3.0	95.88	96.05	96.47	96.13
	3.75 × 3.75	94.99	94.87	97.65	95.84
2.00	1.5 × 1.5	98.76	99.36	98.71	98.94
	2.25 × 2.25	97.47	98.07	97.91	97.82
	3.0 × 3.0	97.42	97.18	96.14	96.91
	3.75 × 3.75	96.53	96.16	96.71	96.47

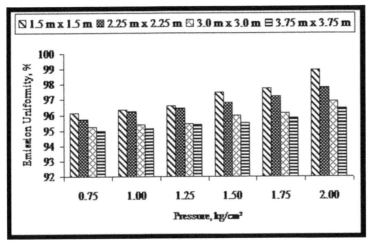

FIGURE 5.23 Emission uniformity as influenced by operating pressures for different spacings of micro sprinkler.

Firake et al.[54] reported the EU of the system ranged from 96.1 to 97% and 96.9 to 97.5%, respectively, for 33 and 57 lph discharge of the micro sprinkler. Shinde et al.[13] have also reported that the average EU of the system was 91.0%. Thus, the findings of this study are in close agreement with those reported by these researchers.

5.4.5 DISTRIBUTION UNIFORMITY

The DU (%) values are indicated in Table 5.11 and Fig. 5.24. For a micro sprinkler spacing of 1.5 × 1.5 m and an operating pressure of 0.75 kg/cm², the estimated value of DU was 71.94%, compared with 73.47% at an operating pressure of 1.0 kg/cm². With further increment in operating pressure, slight improvement in DU was noticed. For the micro sprinkler spacing of 2.25 × 2.25 m and an operating pressure of 0.75 kg/cm², the DU was 68.06%. For the same spacing and at 1.0 kg/cm², the DU improved slightly to 68.15%. Therefore, in general, it was noticed that with increase in operating pressure, there was slight improvement in DU for a given spacing.

TABLE 5.11   Distribution Uniformity (DU) for Different Operating Pressures and Micro Sprinkler Spacings

Operating Pressure (kg/cm²)	Spacing (m)	Distribution Uniformity, DU (%)			Average DU (%)
		I	II	III	
0.75	1.5 × 1.5	70.98	71.20	73.63	71.94
	2.25 × 2.25	64.28	68.56	71.33	68.06
	3.0 × 3.0	44.82	56.54	61.84	54.40
	3.75 × 3.75	17.52	21.00	28.44	22.32
1.00	1.5 × 1.5	71.40	74.10	74.91	73.47
	2.25 × 2.25	74.86	63.97	65.63	68.15
	3.0 × 3.0	67.06	59.10	63.52	63.23
	3.75 × 3.75	36.30	28.10	33.37	32.59
1.25	1.5 × 1.5	74.22	73.65	72.63	73.50
	2.25 × 2.25	69.81	70.24	70.10	70.05
	3.0 × 3.0	64.79	64.46	63.61	64.29
	3.75 × 3.75	34.82	39.83	35.16	36.60
1.50	1.5 × 1.5	75.55	73.20	71.82	73.52
	2.25 × 2.25	69.76	70.58	70.43	70.26
	3.0 × 3.0	66.16	65.64	67.54	66.45
	3.75 × 3.75	32.34	41.99	44.22	39.52
1.75	1.5 × 1.5	76.50	71.92	74.64	74.35
	2.25 × 2.25	75.26	69.20	68.59	71.02
	3.0 × 3.0	69.27	66.66	69.64	68.52
	3.75 × 3.75	43.89	47.29	50.31	47.16
2.00	1.5 × 1.5	71.63	74.68	78.21	74.84
	2.25 × 2.25	73.53	73.16	74.17	73.62
	3.0 × 3.0	73.72	70.71	65.10	69.84
	3.75 × 3.75	49.29	49.36	47.43	48.69

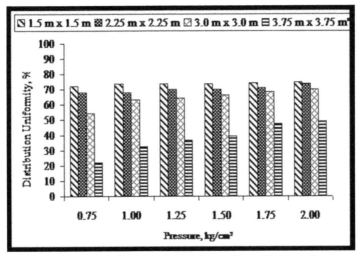

FIGURE 5.24   Distribution uniformity (DU, %) as affected by operating pressures for different spacings of micro sprinkler.

When spacing was changed from 1.5×1.5 m to 2.25×2.25 m, the reduction of DU was 3.88% at an operating pressure of 0.75 kg/cm². Further increment in the spacing to 3.0×3.0 m and 3.75×3.75 m, the reduction was 13.66% and 32.08%, respectively. Thus, it was revealed that the micro sprinkler spacing influenced the value of DU substantially compared with operating pressure. In general, it can be concluded that for higher values of DU, micro sprinklers should be operated comparatively at higher pressures in order to have a desired value of DU. Keller and Merriam[6] have recommended that the value of DU must be greater than 75%. The same criterion was adopted for the study, and sprinklers should be spaced at 1.5×1.5 m and operated at a pressure greater than 1.0 kg/cm² or should be spaced at 2.25×2.25 m and operated at 2.0 kg/cm².

The statistical analysis revealed that the interaction between spacings of micro sprinkler and operating pressures significantly affected the DU (Table 5.12). In the calculation of DU, depth of precipitation is taken into consideration, which is a function of discharge and overlapping percentage. Ultimately, the DU is influenced by both operating pressures and micro sprinkler spacings. The results of DU in Tables 5.11 and 5.12 confirm this presumption. Table 5.12 also indicates that the increase in spacing drastically reduced the DU.

It was observed that the operating pressure was increased with constant spacing, which increases the DU steadily. The maximum DU of 74.84% was achieved at an operating pressure of 2.0 kg/cm² and at a spacing of 1.5×1.5 m.

Results indicate that the relationship was linear between DU for all micro sprinkler spacings and the operating pressure under study, as shown below:

$$DU = C(P) + d \tag{17}$$

The regression coefficients (C and d) are reported in Table 5.13. The value of intercept (d) ranged from 71.88 to 19.96, whereas the slope of line (C) varied from 0.49 to 5.09. The correlation coefficients (R^2) were greater than 0.85.

TABLE 5.12   ANOVA Showing the Effects of Different Operating Pressures and Micro Sprinkler Spacings on Distribution Uniformity (DU, %)

Source	DF	SS	MS	F	SE (m)	CD (5%)	CD (1%)
A = operating pressure	5	1211.24	242.248	18.784[a]	1.04	2.948	3.934
B = micro sprinkler Spacing	3	14252.3	4750.76	368.377[a]	0.8464	2.407	3.212
A × B	15	756.019	50.4012	3.908[a]	2.073	5.896	7.868
Error	48	619.03	12.8965				

[a]Significant at 1%.

Table 5.13 Values of Constant C, d, and Correlation Coefficient (R^2) for Different Spacings

Spacing, m	C	d	R²
1.5 × 1.5	0.4903	71.88	0.86
2.25 × 2.25	1.0463	66.53	0.90
3.0 × 3.0	2.7209	54.93	0.85
3.75 × 3.75	5.0991	19.96	0.95

5.5   CONCLUSIONS

The increasing demand of food for the growing population is enforcing need for expansion of irrigation resources throughout the world. Water being a limited resource, its efficient use is a base to feed ever-increasing world population. This is possible only by the way of better water management practices in the field and by introducing advance methods of irrigation. Pressurized irrigation systems have been introduced to prevent the unavoidable losses and nonuniform distribution of water associated with surface methods. A micro irrigation system has the potential to be a very efficient irrigation method. The basic information needed for designing the efficient micro sprinkler irrigation method is inadequate. Therefore, an attempt was made to generate the information on pressure-discharge relationships, precipitation pattern, UC, EU, and DU for the micro sprinklers that are locally available in the market. The field experiment was conducted to study the effects of micro sprinkler spacings and operating pressures at Instructional Farm of CTAE, Udaipur. The results are summarized as follows:

1. The discharge of micro sprinkler under study was affected by the operating pressure. Maximum discharge of 73.10 lph at an operating pressure of 2.0 kg/cm² and minimum discharge of 32.99 lph at an operating pressure of 0.75 kg/cm² were recorded.

2. Mathematical equation of the form $Q = K P^x$ was developed for the micro sprinkler under study. The characteristic constant (K) was 44.45 and discharge exponent (x) was 0.78 with correlation coefficient of 0.95. The high value of correlation coefficient (R^2) indicated goodness of fit.

3. Precipitation distribution pattern for 1.0, 1.5, 2.0, and 2.5 kg/cm² operating pressure was studied by drawing the precipitation contours at an interval of 1.5 mm. From the precipitation contours, it was observed that contours were closely spaced nearer to the micro sprinkler and widely spaced toward the periphery. The distribution profiles were also drawn and studied to get an idea about the non-uniformity in precipitation distribution. At a lower operating pressure (1.0 kg/cm²), the shape of the distribution profile was found as flat elliptical type, which gradually changed to triangular type with increase in operating pressure. Flat elliptical type of distribution profile is the desired profile, which allows minimum overlapping of adjacent micro sprinklers. Hence, wider spacings between the micro sprinklers keep the total cost of the system on the lower side compared with the micro sprinklers having triangular type of distribution. The latter necessitates closer spacing between them for uniform distribution pattern, thus involving higher cost per unit area.

4. Effective radius ranged from 1.0 to 1.75 m at an operating pressure of 1.0–2.5 kg/cm². Thus, if we consider the overlapping of 50% under low-wind-speed conditions, the spacing between micro sprinkler operated at 1.0 kg/cm² should be 1.0 m, and when the system is operated at 2.5 kg/cm², the maximum spacing with 50% overlapping should be 1.75 m. However, the system at low pressure (1.0 kg/cm²) causes increase in cost, and high pressure (2.5 kg/cm²) may reduce the uniformity of precipitation pattern. Thus, if the system runs between operating pressures of 1.5–2.0 kg/cm², the maximum spacing can vary from 1.25 to 1.5 m with 50% overlapping under low-wind-speed conditions.

5. The minimum average application depth was 7.69 mm/h, and the maximum was 9.23 mm/h.

6. Effective maximum depths were 15.56, 14.14, 14.00, and 14.95 mm/h at 1.0, 1.5, 2.0, and 2.5 kg/cm^2 of operating pressure, respectively. Higher effective maximum depth indicated the concentration of precipitation nearer to the center of the micro sprinkler. Lower value of effective maximum depth indicated nearly uniform distribution of irrigation water from the center to the periphery.

7. Absolute maximum depths of 19.80, 17.54, 17.26, and 21.78 mm/h were recorded at 1.0, 1.5, 2.0, and 2.5 kg/cm^2 of operating pressures, respectively. A lower value of absolute maximum depth of 17.26 mm/h for 2.0 kg/cm^2 is an indicative of superior precipitation distribution patterns over other pressures under study.

8. Mean application depth of 11.40, 8.08, 6.72, and 7.62 mm/h were recorded at 1.0, 1.5, 2.0, and 2.5 kg/cm^2 of operating pressures, respectively.

9. A distribution characteristic is the ratio of area receiving the depth greater than mean depth to the total wetted area. The values were 24.07, 17.20, 18.70, and 36.27% at 1.0, 1.5, 2.0, and 2.5 kg/cm^2 of operating pressures, respectively. These results confirm that the distribution profile is elliptical at lower operating pressure, and it gradually changed to triangular type with increase in operating pressure. The findings also confirm that micro sprinkler under study should be operated in the range of 1.5–2.0 kg/cm^2 pressure.

10. The CV was 0.67, 0.77, 0.86, and 0.83 at operating pressures of 1.0, 1.5, 2.0, and 2.5 kg/cm^2, respectively. A minimum value of CV was 0.67 was at an operating pressure of 1.0 kg/cm^2. Lower CV value indicates the superiority of precipitation pattern corresponding to 1.0 kg/cm^2 operating pressure.

11. The relationship between operating pressure and UC for different micro sprinkler spacings of the form $UC = m(P)^n$ was developed. The values of constants m and n ranged from 33.85 to 80.18 and 0.037 to 0.325, respectively. The values of exponent n indicate the variation of UC with respect to operating pressure. The lower values indicate less variation of UC with operating pressure and vice versa. The characteristic constant m is the highest for the spacing of 1.5 × 1.5 m, indicating at least 80% UC at 1.0 kg/cm^2 pressure, which is desirable for high performance of the system. This confirms that the spacing of 1.5 × 1.5 m is suitable for micro sprinkler as compared with higher spacing.

12. When the operating pressure was changed from 0.75 to 2.0 kg/cm^2 for micro sprinkler spacing of 1.5 × 1.5 m, the EU values changed from 96.13 to 98.94%. Thus, it was observed that the EU did not change substantially either due to operating pressures or due to micro sprinkler spacing.

13. For a micro sprinkler spacing of 1.5 × 1.5 m and an operating pressure of 0.75 kg/cm^2, the DU was 71.94%. For the same operating pressure when the spacing were increased to 2.25 × 2.25 m, 3.0 × 3.0 m, and 3.75 × 3.75 m, the values of DU were 68.06, 54.40, and 22.32%, respectively. The straight line,

DU = C P + d, was fitted to establish the functional relationship between the DU and the operating pressure. The values of the intercept (d) ranged from 71.88 to 19.96, whereas the slope of line (c) ranged from 0.49 to 5.09. The correlation coefficient (R^2) was higher than 0.85. Higher value of R^2 indicates that there was a goodness of fit.

Following conclusions are drawn from the present investigation:

1. Pressure–discharge relationship followed a power law: $Q = K P^x$. The equation indicates that discharge was increased with increase in the operating pressure. However, this increase was stabilized after the pressure exceeded to 1.5 kg/cm².

2. Precipitation pattern was flat elliptical type at low operating pressure, and it gradually changed to triangular type with an increase in pressure. The precipitation performance parameter indicates that the micro sprinkler should be operated at 1.5–2.0 kg/cm² under low-wind-speed conditions.

3. The relationship between the UC and the operating pressure for different spacings was of power type: UC = m p^n.

4. The UC, EU, and DU values indicate that the micro sprinkler should be operated at a pressure range of 1.5–2.0 kg/cm², and suitable spacing for this pressure range was 1.5 × 1.5 m.

5. The savings in the number of sprinklers per hectare were 55 and 36% with the spacing of 1.5 × 1.5 m and 1.25 × 1.25 m, respectively, compared with 1.0 × 1.0 m spacing. Similarly, the savings in lateral length/ha were 33 and 20% with these spacings compared with a spacing of 1.0 1.0 m.

6. The relationship between DU and operating pressure was linear: DU = C P + d.

5.6 SUMMARY

The performance characteristics of micro sprinkler include pressure discharge relationship, precipitation pattern, UC, EU, and DU. These characteristics have been evaluated under various operating conditions. Suitable spacings between micro sprinklers were also tested at an operating pressure ranging from 0.75 to 2.0 kg/cm². The experiments were conducted for the pressure ranges from 1.0 to 2.5 kg/cm² for precipitation pattern treatment and micro sprinkler spacing of 1.5× 1.5 m, 2.25× 2.25 m, 3.0 × 3.0 m, and 3.75 × 3.75 m under low-wind-speed condition. Pressure–discharge relationship followed the power law: $Q = 44.45 P^{0.78}$. Precipitation pattern was found to be flat elliptical type at low operating pressure, and it gradually changed to triangular type with increase in pressure. The EU was > 90%. The UC and DU increased with increasing operating pressure and decreased with increase in micro sprinkler spacing. It was concluded that the micro sprinklers should be operated at a pressure range of 1.5–2.0 kg/cm² and at a spacing of 1.5 × 1.5 m.

KEY WORDS

- **American Society of Agricultural Engineers**
- **discharge**
- **distribution uniformity**
- **emission uniformity**
- **emitter**
- **field performance**
- **India**
- **Indian Society of Agricultural Engineers**
- **irrigation throw diameter**
- **micro irrigation**
- **micro sprinkler**
- **precipitation pattern**
- **pressure**
- **pressure–discharge relationship**
- **spray jet**
- **spray pattern**
- **sprinkler spacing**
- **uniformity coefficient**
- **uniformity of water application**

REFERENCES

1. Shivanappan, R. K. *The Hindu Survey of Indian Agriculture*; M/S Kasturi and Sons Ltd., National Press: Chennai, India, 2000.
2. MIBC. *India – 2005: A Reference Annual*; Government of India (Agriculture Section), 2005; pp 64–75.
3. Mane, M. S.; Ayare, B. L. *Principles of Sprinkler Irrigation*; Jain Brothers Publication, 2007; pp 93–122.
4. Christiansen, J. E. Irrigation by Sprinkling. In California Agric. Exp. Stn. Bull. 670; University of California: Berkeley, 1942.
5. Seginer, I. Water distribution from a medium pressure sprinkler. *J. Irrig. Drain. Div. (ASCE)*, 1963, 89(IR-II), 13–19.

6. Keller, J.; Karmeli, D. Trickle irrigation design parameters. *Trans. Am. Soc. Agric. Eng.* 1974, 17, 678–684.
7. Voigt, D. Effect of design on hydraulic performance of rotary sprinkler. *Agric. Eng. Abstr.* 1980, 5, 14.
8. Giari, M.; Rossi, N.; Taglioli, G. Discharge rate tests on dripper and sprinkler emitters. *Agric. Eng. Abstr.* 1982, 7(4), 104.
9. Singh, A. K.; Jain, S.; Mathur, L. N. Micro Sprinkler Performance Evaluation and Constraints for its Adoption. In Proceeding XI International Congress on Use of Plastics in Agriculture Held at New Delhi, 1990; pp 79–84.
10. Firake, N. N.; Salunkhe, D. S. Soil moisture movement in micro sprinkler irrigation. *Agric. Eng. Today* 1992, 15(1–6), 52–55.
11. Sakore, D. K. Hydraulics of Micro Sprinkler. Unpublished B. Tech. (Agril. Eng.) Thesis, Mahatma Phule Krishi Vidyapeeth, Rahuri, 1992.
12. Firake, N. N.; Salunkhe, D. S.; Pampattiwar, P. S., Evaluation of hydraulic performance of micro sprinkler irrigation system. *Indian J. Agric. Eng.* 1992, 1(2), 141–144.
13. Shinde, R. U.; Darde, S. R.; Firake N. N. Field Evaluation of Hydraulic Performance of Static Micro Sprinkler Irrigation System. Unpublished B. Tech. (Agril. Eng.) Thesis, Mahatma Phule Krishi Vidyapeeth, Rahuri, 1993.
14. Singh, B. Performance Evaluation of Micro Sprinkler. Unpublished M.E. (Agril. Eng.) Thesis, Rajasthan Agriculture University, Bikaner, 1993.
15. Gawali, V. M.; Budhan K. M. Hydraulic Performance of Micro Sprinkler. Unpublished B.Tech. (Agril. Eng.) Thesis, Mahatma Phule Krishi Vidyapeeth, Rahuri (M.S.), 1994.
16. Lonkar, S. B.; Dhage, S. T. Studies on Hydraulic Performance of Micro Sprinkler System. Unpublished B. Tech. (Agril. Eng.) Thesis, Mahatma Phule Krishi Vidyapeeth, Rahuri, 1998.
17. Suryawanshi, S. C.; Kolte, M. M.; Patil, S. R. Hydraulic Characteristics of Micro Sprinkler. Unpublished B. Tech (Agril. Eng.) Thesis, Mahatma Phule Krishi Vidyapeeth, Rahuri (M.S.), 1999.
18. Singh, R.; Kale, M. U.; Chandra, A. Performance Evaluation of Micro Jets. In Proceedings of International Conference on Micro and Sprinkler Irrigation Systems, February 8-10; Jalgaon, 2001; pp 155–162.
19. Pampattiwar, P. S.; Kadam, U. S.; Kadam, S. A.; Sharma, P. S. Hydraulic Performance of Micro Sprinklers. In XXXVI Annual Convention of Indian Society of Agricultural Engineering; IIT: Kharagpur, 2002.
20. Patil, P. V.; Garje, U. K.; Khachane, S. V. Studied on the Hydraulic Performance Evaluation of Different Types of Micro Sprinklers. Unpublished M. Tech. (Agril. Eng.) Thesis, Mahatma Phule Krishi Vidyapeeth, Rahuri, 2002.
21. Barragan, J.; Wu, I. P. Simple pressure parameters for micro irrigation design. *J. Biosyst. Eng.* 2005, 90(4), 463–475.
22. Kumar, J. Operating Characteristics Analysis of Rotary Sprinkler Irrigation System. Unpublished M.E. (SWCE) Thesis, Maharana Pratap University of Agriculture and Technology, Udaipur, 2007.
23. Fisher, G. R.; Wallender, W. W. Collector size and test duration effect on sprinkler water distribution measurement. *Trans. Am. Soc. Agric. Eng.* 1988, 31, 538–542.
24. Madramootoo, C. A.; Khatri, K. C.; Rigby, M. Hydraulic performance of five different trickle irrigation emitters. *Can. Agric. Eng.* 1988, 30(1–2), 1–4.
25. Boman, J. B. Distribution pattern of micro irrigation spinner and spray emitters. *Appl. Eng. Agric.* 1989, 5(1), 50–55.
26. Gutal, G. B.; Chougale, A. A.; Kulkarni, P. V. Comparative Study of Drip, Micro Sprinkler, Biwall and Border Irrigation on Groundnut. In Annual Report of Research Review Subcommittee on Groundnut (IDE); Mahatma Phule Krishi Vidyapeeth: Rahuri, 1989; pp 17–20.

27. Pathare, S. B. Design of Micro Sprinkler system Based on Non-Uniformity Sprinkling. Unpublished M. Tech. (Agril.Eng.) Thesis, Mahatma Phule Krishi Vidyapeeth, Rahuri, 1993.

28. Aragade, K. B.; Thombal, N. V. Studies on Pressure Discharge Throw Diameter Relationship and Spray Pattern of Micro-Sprinkler. Unpublished B. Tech. (Agril. Eng.) Thesis, Mahatma Phule Krishi Vidyapeeth, Rahuri, 1994.

29. Pandey, A. K., Chaudhary, H. S., Dukla, K. N.; Singh, K. F. Micro Sprinkler Irrigation for Changing World. Proceeding of the Fifth International Congress Held During, April 2-6; Hyat Regency: Orlando, Florida, 1995; pp 857–862.

30. Keller, J.; Merriam, J. L. *Farm Irrigation System Evaluation: A Guide for Management*; Agricultural and Irrigation Engineering Department, Utah State University: Logan, Utah, 1978; p 255.

31. Shete, D. T. and Modi, P. M, Sprinkler Performance Characteristics with Respect to Radial and Grid Catch Can Patterns. In ICID, Water Resources Eng. and M.S. University of Baroda, Samiala, Dist. Vadodara, Gujrat, 391410, India, 1995.

32. Vishnu, B. L. J.; Santhana, B. S. Performance evaluation of sprinkler and spray type micro-irrigation emitters. *J. Water Manage.* 1995, 3(1–2), 1–3.

33. Mateos, L. Assessing whole field uniformity of stationary sprinkler irrigation systems. *J. Irrig. Sci.* 1998, 18, 215–216.

34. Singh, R. P.; Singh, V. P.; Naipali, B. Drop Size Distribution, Water Application Profile and Uniformity of Micro Sprinkler. In Proceeding of Workshop in Micro Irrigation and Sprinkler Irrigation Systems, April 28-30; New Delhi, Paper # III-23, 1998.

35. DeBoer, W. D. Drop and energy characteristics of a rotating spray plate sprinkler. *J. Irrig. Drain. Eng. Am. Soc. Civ. Eng.* 2002, 128, 137–146.

36. Faci, J. M.; Salvador, R.; Playan, E. Comparison of fixed and rotating spray plate sprinklers. *J. Irrig. Drain. Eng. Am. Soc. Civ. Eng.* 2001, 127, 224–233.

37. DeBoer, W. D. Sprinkler application pattern shape and surface runoff. *Trans. Am. Soc. Agric. Eng.* 2001, 44, 1217–1220.

38. Clark, G. A.; Srinivas, K.; Rogers, D. H.; Stratton, R.; Martin, V. L. Measured and simulate uniformity of low drift nozzle sprinklers. *Trans. Am. Soc. Agric. Eng.* 2003, 46, 321–330.

39. Dogan, E.; Clark, G. A.; Rogers, D. H.; Martin, V. L. Evaluation of Collector Size for the Measurement of Irrigation Depths. In American Society of Agricultural Engineering Annual Meeting, Paper # 032007, 2003.

40. Nehete, P. M.; Kulkarni, S. S.; Kuchekar, V. G. Studies on Hydraulic Performance Evaluation of Micro Sprinkles. Unpublished B.Tech. (Agril.Eng.) Thesis, Mahatma Phule Krishi Vidyapeeth, Rahuri (M.S.), 2003.

41. Sourell, H.; Faci, J.; Playan, E. Performance of rotating spray plate sprinklers in indoor experiments. *J. Irrig. Drain. Eng. Am. Soc. Civ. Eng.* 2003, 129, 376–380.

42. Playan, E.; Garrido, S.; Faci, J. M.; Galán, A. Characterizing pivot sprinklers using an experimental irrigation machine. *Agric. Water Manage.* 2004, 70, 177–193.

43. James, B. K.; Mal, B. C.; Tiwari, K. N. A Method for Determination of Optimum Design Parameters of a Micro Sprinkler or Micro-Jet. In 40th Annual Convention and Symposium of Indian Society of Agricultural Engineering, January 19-21; Tamil Nadu Agriculture University: Coimbatore, Paper # 4.24, 2006.

44. Poul, J. C.; Mishra, J. N.; Pradhan, P. C. Comparative Study of Different Micro Irrigation Systems in Tuber Rose. In 40th Annual Convention and Symposium of Indian Society of Agricultural Engineering, January 19-21; Tamil Nadu Agriculture University: Coimbatore, Paper # 4.26, 2006.

45. Kadam, S. A.; Gorantiwar, S. D.; Kadam, U. S. Study of Hydraulic Performance of Micro Sprinklers. In XLII Indian Society of Agricultural Engineering Annual Convention and Symposium, February 1-3; CIAE: Bhopal, 2008.

46. Keller, J.; Bliesner, R. D. *Sprinkler and Trickle Irrigation*; Van Strand Reinold: New York, 1990. ISBN-O-4442-24645-5.
47. Solomon, K. Variability of sprinkler coefficient of uniformity test results. *Trans. Am. Soc. Agric. Eng.* 1979, 22, 1078–1080.
48. Forkel, H.; Mirshei, W. A new formula for calculating the range of rotating sprinkler. *Agric. Eng. Abstr.* 1980, 5, 41.
49. Ricardo, A. L.; Brito, A.; Willardson, L. S. Sprinkler irrigation uniformity requirement for the elimination of leaching. *Trans. Am. Soc. Agric. Eng.* 1982, 1258–1261.
50. Post, S. E. C., Peek, D. E., Brendler, B. A., Sakovich, N. J. and Waddel, L. Evaluation of Non-Overlapping, Low Flow Sprinklers. In: Drip/Trickle irrigation in Action. Proceeding of the Third Congress Michigan, Vol. 1; California, l985; pp 293–305.
51. Hills, D. I.; Silveirn, R. E. M.; Wallender, W. W. Oscillating pressure for improving application uniformity spray emitters. *Trans. Am. Soc. Agric. Eng.* 1986, 29(4), 1080–1085.
52. Gutal, G. B.; Chougale, A. A.; Kulkarni, P. V. Comparative Study of Drip and Micro Sprinkler Irrigation on Groundnut. In Annual Report of Research Review Subcommittee on Groundnut (IDE); Mahatma Phule Krishi Vidyapeeth: Rahuri, 1988; pp 15–21.
53. Sharma, R. K.; Batawar, H. B. Performance of Sprinkler Irrigation System. In Paper Presented at Silver Jubilee Convention of Indian Society Agricultural Engineering Held at CTAE, Udaipur During, January 5-7; 1989; pp 48–51.
54. Firake, N. N.; Salunkhe, D. S.; Magar, S. S. Effect of uniformity coefficient of water application on spacing between micro sprinkler in vegetables. *Maharashtra J. Hortic.* 1993, 7(1), 38–41.
55. Buzescu, D. M.; Popa, G. Reconsideration of micro sprinklers as a method of vegetable crops. *Angle Institutel de CertariPenstruLegumic Cultura ScFloriculturaVidya*, 1996, 14, 324–330.
56. Hills, D. J.; Barragan, J. Application uniformity for fixed and rotating spray plate sprinkler. *Trans. Am. Soc. Agric. Eng.* 1997, 14, 33–36.
57. Tarjuelo, J. M.; Montero, J.; Honrubia, F. T.; Ortizand, J. J.; Ortego, J. F. Analysis of uniformity of sprinkler irrigation in semi-arid area. *Agric. Water Manage.* 1999, 315–331.
58. Schneider, A. D. Application efficiency and uniformity coefficients for LEPA and spray sprinkler irrigation methods: a review. *Trans. Am. Soc. Agric. Eng.* 2000, 43, 937–944.
59. Rocha, F. A.; Cesar, J. H. F.; Mello, C.R.; Pereira, G. M. Hydraulic performance and water distribution profile of two models of the supermamkad mini-sprinkler. *Rev. Bras. Ing. Agric. Ambiental*, 2001, 5(3), 386–390.
60. Burt, C. M. Rapid field evaluation of drip and micro spray distribution uniformity. *J. Irrig. Drain. Syst.* 2004, 18(4), 275–297.
61. Melissa, C. B.; Michael, D.; Dukes, P. E.; Grady, L. M. Analysis of residential irrigation distribution uniformity. *J. Irrig. Drain. Eng.* 2005, 131, 336–341.
62. Barragan, J.; Bralts, V.; Wu, I. P. Assessment of emission uniformity for micro irrigation design. *J. Biosyst. Eng.* 2006, 93(1), 89–87.
63. Bansod, R. D.; Shukla, K. N. Evaluation of Micro Sprinklers for Uniformity of Water Application. In 40th Annual Convention and Symposium of Indian Society of Agricultural Engineers, January 19-21; Tamil Nadu Agricultural University: Coimbatore, 2006.
64. Kishore, R.; Singh, V. V.; Yadav, B. G. Water Management with Pressure Compensated Micro Sprinklers for Precision Farming. In 40th Annual Convention and Symposium of Indian Society of Agricultural Engineering, January 19-21; Tamil Nadu Agriculture University: Coimbatore, Paper # 4.36, 2006.
65. Jadhav, S. B.; Bhagyawant, R. G.; Pawar, S. N. Hydraulic Performance of Micro Sprinkler Irrigation. In 41st Annual Convention and Symposium of Indian Society of Agricultural Engineering, January 29-31; CAET – JAU: Junagad, Paper # 2.55, 2007.

66. Michael, A. M. *Irrigation – Theory and Practice*. Vikas Publishing House Pvt. Ltd.: New Delhi, 1978; p 644.
67. Walker, W. R. Explicit sprinkler irrigation uniformity, efficiency model. *J. Irrig. Drain. Eng. (ASCE)* 1979, 105, 129–136.
68. Patil, P. S. Studies the Hydraulic Performance Evaluation of Different Types of Micro Sprinklers. Unpublished B.Tech (Agril. Eng.) Thesis, Mahatma Phule Krishi Vidyapeeth, Rahuri, 2002.

APPENDIX I: SAMPLE EXAMPLE

Calculations to evaluate the performance parameter of micro sprinkler at an operating pressure of 1.0 kg/cm^2 by Keller method.

R = Wetted radius = 1.2 m

First concentric radius = $R \times 0.40$ = 1.2 × 0.40 = 0.48 m

Second concentric radius = $R \times 0.60$ = 1.2 × 0.60 = 0.72 m

Third concentric radius = $R \times 0.78$ = 1.2 × 0.78 = 0.94 m

Fourth concentric radius = $R \times 0.93$ = 1.2 × 0.93 = 1.12 m

- Average precipitation depth d_1 in the first concentric ring

$$d_1 = \frac{8.83 + 16.91 + 14.99 + 13.92}{4} = \frac{54.65}{4} = 13.66 \text{ mm/h}$$

- Average precipitation depth d_2 in the second concentric ring
  $$d_2 = 0.0 \text{ mm/h}$$
- Average precipitation depth d_3 in the third concentric ring

$$d_3 = \frac{17.26 + 14.99 + 17.83 + 12.17 + 8.71 + 14.99 + 12.73 + 19.80}{8} = \frac{118.48}{8} \text{ or}$$

$d_3 = 14.81$ mm/h

- Average precipitation depth d_4 in the fourth concentric ring

$$d_4 = \frac{0.51 + 2.66 + 0.86 + 5.15}{4} \lim_{x \to \infty} = \frac{9.18}{4} = 2.29 \text{ mm/h}$$

(1) Average application depth

$$d = \frac{d_1 + d_2 + d_3 + d_4}{4} = \frac{13.66 + 0 + 14.81 + 2.29}{4} = \frac{30.76}{4} = 7.69 \text{ mm/h}$$

(2) Effective maximum depth (mm/h)

$$= \frac{16.91 + 14.99 + 13.92 + 17.26 + 14.99 + 17.83 + 12.17 + 14.99 + 12.73 + 19.80}{10}$$

$$= \frac{155.59}{10} = 15.56 \text{ mm/h}$$

(3) Absolute maximum depth = 19.80 mm/h

(4) Mean application depth

$$= \frac{\sum d}{N} = \frac{188.126 - 5.82}{23 - 7} = \frac{182.30}{16} = 11.40 \text{ mm/h}$$

(5) Precipitation pattern as distribution characteristics

$$\text{Distribution Characteristic} = \frac{\text{Area receiving the depth greater than average depth}}{\text{Total wetted area}} \times 100$$

$$= \frac{1.09}{4.52} \times 100$$

$$= 24.07\%$$

(6) Coefficient of variation

$$\sigma = \sqrt{\frac{\sum di^2}{N}} \text{ and}$$

$$\sigma = \sqrt{\frac{\sum di^2}{N}} \text{ or } \sigma = \sqrt{\frac{\sum 1367.30}{23}} = 7.71 \text{ and}$$

$$CV = \frac{7.71}{11.40} = 0.67$$

(7) Effective radius, Re = 1.0 m
Coefficient of variation for 1.0 kg/cm² by Keller method

d	da	$di = (d-da)$	di^2
0.45	11.40	−10.95	119.90
0.51	11.40	−10.89	118.59
17.26	11.40	5.86	34.34
14.99	11.40	3.59	12.89
2.66	11.40	−8.74	76.39
1.30	11.40	−10.10	102.01

19.80	11.40	8.40	70.56
8.83	11.40	−2.57	6.60
16.91	11.40	5.51	30.36
17.83	11.40	6.43	41.34
0.10	11.40	−11.30	127.78
1.80	11.40	−9.60	92.16
12.73	11.40	1.33	1.77
13.92	11.40	2.52	6.35
14.99	11.40	3.59	12.89
12.17	11.40	0.77	0.59
0.34	11.40	−11.06	122.32
5.15	11.40	−6.25	39.06
14.99	11.40	3.59	12.89
8.71	11.40	−2.69	7.24
0.86	11.40	−10.54	111.09
1.30	11.40	−10.10	102.01
0.53	11.40	−10.87	118.16
$\sum$ 188.13			$\sum$ 1367.30
Total observation = 23			

Observations to be considered for the estimation of average application depth by concentric ring method given by Keller for micro sprinkler at 1.0 kg/cm²

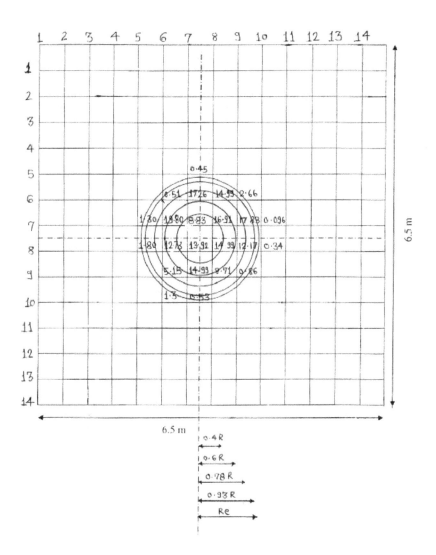

APPENDIX II: SAMPLE EXAMPLE

Observation of precipitation depth (mm) for 30 min at grid points for calculating uniformity coefficient and distribution uniformity for 1.5 × 1.5 m spacing at 0.75 kg/cm²

R I

S1	7.92	9.05	8.18	6.68	S4	5.37	6.79	7.52	5.85	S7
6.22	7.64	9.62	8.40	7.74	5.09	5.66	6.79	8.49	6.51	5.94
8.77	9.62	7.64	7.52	7.30	9.62	6.79	8.20	7.64	7.30	10.19
7.92	7.36	7.64	9.29	7.07	6.19	7.32	9.05	7.52	7.36	7.92
5.66	4.53	6.79	5.75	4.86	5.09	4.24	6.22	9.73	6.11	6.79
S2	5.30	8.20	9.05	4.42	S5	5.38	7.92	9.62	4.81	S8
7.36	5.66	7.07	4.24	4.81	6.22	4.86	11.32	8.20	5.09	6.19
8.49	7.07	9.05	8.49	6.22	8.20	7.36	9.62	10.19	5.67	9.05
9.62	6.79	9.90	6.79	9.05	7.36	7.07	8.49	9.62	5.66	9.62
7.74	6.51	7.36	4.53	4.53	5.94	4.96	8.49	9.62	4.81	5.09
S3	6.22	6.19	4.89	5.09	S6	6.51	7.36	8.49	6.79	S9
R II										
S1	6.19	9.51	10.84	5.75	S4	6.19	10.17	8.40	4.42	S7
6.63	5.30	9.29	9.07	5.30	5.81	6.63	10.84	8.18	4.89	6.63
7.52	9.90	10.84	10.10	4.89	10.17	4.53	9.94	9.51	8.40	7.07
8.40	8.84	10.17	10.10	5.30	7.52	5.30	9.62	9.29	7.07	6.63
7.07	6.63	7.07	7.96	5.38	6.63	5.75	9.07	8.84	5.66	5.85
S2	4.89	7.52	9.51	5.85	S5	6.79	8.49	7.92	6.51	S8
4.86	5.75	7.52	7.07	5.30	5.75	6.81	9.51	8.40	5.85	6.19
7.36	10.10	6.19	6.19	4.85	7.96	6.11	7.64	7.64	7.52	7.74
8.49	9.51	7.07	7.96	5.97	9.73	6.22	7.36	7.52	7.36	6.19
5.75	5.97	8.84	9.73	6.63	6.63	6.19	7.07	7.74	5.97	5.97
S3	4.42	10.84	8.40	5.75	S6	5.75	7.36	5.37	6.63	S9
R III										
S1	6.11	8.20	9.05	6.22	S4	7.36	8.49	8.18	6.51	S7
5.38	6.19	7.52	6.79	6.19	5.75	5.66	7.52	8.18	6.22	5.85
7.52	6.22	9.07	7.92	6.51	8.40	7.96	7.74	7.74	8.49	6.22
6.63	6.51	9.29	7.36	5.97	7.96	8.18	7.52	7.52	7.74	7.07
4.89	4.86	8.49	7.30	6.63	6.19	5.97	7.07	8.49	6.79	6.11

S2	5.75	9.51	10.84	6.19	**S5**	5.85	7.36	8.49	6.63	**S8**
5.75	5.75	10.10	10.17	6.63	6.63	5.55	9.05	9.62	6.51	4.53
8.40	5.30	8.84	9.94	4.85	7.07	9.51	8.20	9.73	7.07	7.52
10.17	4.89	7.64	7.07	5.85	7.74	8.84	7.64	9.29	7.36	6.79
5.75	4.53	7.07	7.52	6.22	7.64	7.07	7.07	8.49	5.97	5.30
S3	5.97	6.63	8.84	6.11	**S6**	7.07	7.96	8.40	6.19	S9

CHAPTER 6

EFFICIENCY OF ACIDIFICATION FOR UNCLOGGING OF EMITTERS*

PAWAR PRAMAD SURESH and S. J. SUPEKAR

CONTENTS

*Edited and abbreviated version of "Pawar P. Suresh, July 2011. Studies on efficiency of acids for un-clogging of emitters. M. Tech. Thesis for Department of Irrigation and Drainage Engineering, CAET, Marathwada Krishi Vidyapeeth, Parbhani – 431402 (M.S.), India."

6.1   INTRODUCTION

Irrigation system affects yield and quality of agricultural produce from greenhouse farming system. Therefore, irrigation is a vital and important part of the greenhouse agriculture. India ranks 23rd in international flower export (0.38% of world's floriculture export). In India, area under flowers is estimated to be more than 0.1 million hectare. The total production of flowers in India is 36,568.5 million of loose flowers and 612,152.3 million of cut flowers.[1]

During 2007, annual world production of capsicum amounted to 27.46 million metric tons from an area of 1.72 million hectare. China is the major producer of capsicum (bell pepper) and contributes 36% of the world's cultivated area with a production of 12.53 million tons.[2] India has an average annual production of 0.9 million tons from an area of 0.885 million hectare with a productivity of 1.017 t/ha.[3]

In greenhouse cultivation, it is possible to control the weather parameters suitable for optimal crop growth. At the same time, it provides an environment that maximizes the working efficiency and optimal use of natural resources. The main advantage of greenhouse farming is that production can be obtained throughout the year even under adverse climatic conditions.[4]

Out of 172.6 million hectares of cropped area, only 76.82 million hectares is under irrigation in India. It implies that 44.51% of the cropped area is under irrigation. In Maharashtra, out of 30.8 million hectares of geographical area, 22.5 million hectares is under irrigation, and 0.541 million hectares is under drip irrigation. In Maharashtra, in majority of the areas under drip irrigation, the source of irrigation water is groundwater. In different parts of Maharashtra state, average salt concentration of well water ranges between 425 and 2135 ppm, electrical conductivity (EC) is in the range of 0.66–3.337 dS/m, and pH is in the range of 7.5–8.5.[5] Predominant soluble salts consist of sulfate, magnesium, and sodium. Comparatively higher salt concentration of groundwater is known to cause partial or total clogging of emitters.

The main hurdle in drip irrigation management is emitter clogging. The phenomenon of emitter clogging has been extensively studied by various investigators.[6-9] The emitter cogging can be caused by physical, chemical, and biological agents.[10] Physical clogging is caused by suspended inorganic particles (sand, silt, clay, and plastics), organic materials (animal residues, snails, etc.) and microbiological debris (algae, protozoa, etc.), and physical materials often combined with bacterial slimes. Chemical clogging is due to dissolved solids integrating with each other to form precipitates, such as the precipitation of calcium carbonate in water that is rich in calcium and bicarbonates.[11] Biological clogging is due to algae, iron slimes, and sulfur slimes (Table 6.1).

TABLE 6.1 Criteria for Plugging Potential of Drip Irrigation System

Clogging Agent	Slight	Moderate	Severe
	Parts Per Million (ppm) Except pH		
Physical			
Suspended solids	<50	50–100	>100
Chemical			
pH	<7.0	7.0–7.5	>7.5
Dissolved solids	<500	500–2000	>2000
Manganese	<0.1	0.1–1.5	>1.5
Iron	<0.1	0.1–1.5	>1.5
Hardness	<150	150–300	>300
Hydrogen sulfide	<0.5	0.5–2.0	>2.0
Biological			
Bacteria	<10,000	10,000–50,000	>50,000

To prevent emitter clogging, different methods are in use. Filtering and flushing of drip lines are simple and useful methods to prevent emitter clogging, particularly in case of physical clogging.[12] Filtering can prevent inorganic particles and organic materials suspended in water from entering into the drip irrigation system. Flushing of drip lines can wash out inorganic and organic materials precipitated in emitter orifices and on the inside wall of drip hoses of the drip system. Chemical clogging can be controlled with chlorination or acid injection, which can lower the pH value of irrigation water and thus prevent chemical precipitation. Biological clogging is quite difficult to control. Chlorination is one of the most common and efficient methods to prevent and treat emitter clogging caused by algae and bacteria.[13,14]

Acidification is the injection of acid into drip irrigation system to reduce emitter clogging and is primarily carried out to lower the pH of the irrigation water and to prevent precipitations of salts. Precipitation of salts (calcium carbonate, magnesium carbonate, or ferric oxide) can cause either partial or complete blockage of the system. Acid may also be effective in cleaning the system, which is already partially blocked due to precipitation of salts. The most reliable step for deciding on an acid treatment is the water analysis. Water samples are collected during the survey and then analyzed to recommend acid treatment based on the water quality. Generally, hydrochloric acid, nitric acid, sulfuric acid, and phosphoric acid are used for acid treatment.[15] Argus nutrient dosing handbook[16] is an excellent source and guide for nutrient dosing.

Emitter clogging greatly reduces water distribution uniformity in the irrigated field, which negatively influences crop growth and yield.[17-19]

Considering these factors, project work entitled *Studies on efficiency of acids for unclogging of emitters* was undertaken with the following objectives:

1.   To study efficiency of acidification of trickle irrigation system.
2.   To monitor emission uniformity (EU) for different acid treatments.
3.   To suggest a suitable acid treatment for unclogging of emitters.
4.   To study the response of rose flower and capsicum (bell pepper) crops to different acid treatments.

6.2   REVIEW OF LITERATURE

6.2.1   *SCOPE OF MICRO IRRIGATION*

Evans and Probsting[20] reported that adequate water was available without deep percolation to orchard crops when irrigated at 100% evapotranspiration, under drip irrigation.

Pampattiwar[21] revealed that fruit yield of lime under minimum water stress irrigation treatment was significantly superior over maximum water stress under surface irrigation treatment. The yield of drip-irrigated lime was significantly superior over furrow irrigation.

Modi and Sood[22] indicated that water demand for agriculture will increase from 470 km^3 in 1985 to 740 km^3 in 2025, whereas demand for nonagricultural activities will increase from 70 to 280 km^3. Irrigation is a necessary condition for increasing agricultural production and productivity. Under drip irrigation system, water losses due to conveyance, distribution, and evaporation are reduced to a large extent, and water use efficiency (WUE) is as high as 95% compared with conventional irrigation.

For drip irrigation, William[23] mentioned advantages, disadvantages, system components, and benefits in raising fruit and vegetable crops because of improvement in quality. Narayanamoorthy[24] found micro irrigation to be an efficient irrigation method in saving water and increasing WUE compared with conventional surface irrigation, where WUE is only about 30–40%. Oza[25] has discussed topics such as irrigation sector terminology, what is irrigation, status of irrigation, irrigation and water resources in India, importance of irrigation in the Indian economy, and beneficiaries of irrigation.

Awulachew and Talu[26] stated that drip irrigation supplies water to the soil in the vicinity of plants at low flow rates (0.5–10 lph, depending on emitter type) from lateral line fitted with emitters. Jain (2010) reported that presently more than 0.26 million hectare is micro irrigated in the country. Maharashtra is the state leading in the adoption of micro irrigation. Here, out of the total irrigated area of 2.52 million hectare, 62% of area is irrigated from more than a million privately owned dug wells.

Sai[27] reported that use of drip irrigation enhances irrigation efficiency compared with conventional irrigation. With proper management, application efficiency of a drip irrigation system can range from 80 to 90% for a well-designed, installed, and maintained drip irrigation system compared with 55% for a drip irrigation system without proper management. Drip irrigation can reduce exposure to water risks and input costs making agribusiness operation more resilient, profitable, and solvent.

6.2.2 DRIP IRRIGATION SYSTEM

David and Mills[28] stated that water is becoming more expensive nowadays because of which drip irrigation has increased in popularity. The rise in popularity has coincided with the development of new irrigation technology, combined with a revised philosophy on the part of home gardeners and the green industry to conserve water.

For irrigation planning of a project, Toth[29] reported that the required and utilizable water has to be available with 80% of security. Water resource can be underground water, artificial channels, or natural rivers, lakes, or sewage water.

Anonymus[15] and Jain Irrigation have discussed components of drip irrigation system, types of micro irrigation, advantages and disadvantages of drip irrigation, principle of drip irrigation, and automation of irrigation system. Rogers et al.[30] stated that subsurface drip irrigation can function without all of the listed components; however, it may be difficult to manage and maintain and may perform poorly. Usually, there are several types of each component. Drip irrigation system can be arranged in different layouts. Variations in pressure within pipes will affect the output of individual emitters.

Arizona Landscape Irrigation[31] indicated that components of a drip irrigation system include irrigation controller or timer, backflow preventer, valves, filters, pressure regulator, pipes, micro tubing, emitters, flushing valve, and cap. Knowledge of different components of drip irrigation is essential in design. Poly pipes and pipe fittings should be made by the same manufacturer. The hole puncher should be made by the same manufacturer of the *emitter*.[32-34]

6.2.3 WATER QUALITY IN DRIP IRRIGATION

Patil et al.[35] studied water quality of 25 dug wells during rainy, winter, and summer seasons for irrigation suitability. They observed that well water was alkaline in reaction, and the salinity was increased during the summer season. Clark and Rogers[36] reported that water for drip irrigation can come from wells, ponds, rivers, lakes, municipal water systems, or plastic-lined pits. Water from these various sources will have large differences in quality. Well water and municipal water are generally clean and may require only a screen or disk filter to remove suspensions.

Schultheis[37] reported that water should be analyzed for inorganic solids (sand and silt), organic solids (algae, bacteria, and slime), dissolved solids (iron, manganese, sulfates, chlorides, and carbonates (calcium)), pH, and hardness of water. Table 6.1 indicates effects of physical, chemical, and biological parameters on the clogging potential of drip irrigation systems. Lamont and Orzolek[38,39] reported that water source and water quality analysis should identify inorganic solids, organic solids, and dissolved solids. Rogers et al.[30] reported that recommended water quality tests include electrical conductivity (EC), pH, cations, anions, sodium absorption ratio (SAR), nitrate nitrogen, iron, manganese, hydrogen sulfide, total suspended solids (TSS), bacterial population, boron, and the presence of oil.

Dehghanisanji et al.[13] found that pressure-compensating emitters have smaller variations in emitter discharge than non-pressure-compensating emitters, when low quality of water was induced with algae and protozoa. Also emitter with self-flushing and pressure-compensating emitters should have priority for use in drip irrigation under saline water.

Thokal[40] analyzed the water samples collected from Katepurna, command area of Vidarbha, India. Overall, 90% water samples collected from wells were not fit for irrigation. By use of poor quality water, soil alkalinity and salinity problems occur.

Shrivastava[41] collected water samples from Nagpur main drain orchard region of central India, and all samples were analyzed for salinity indices (total soluble salts, chloride content, and sodality index). They observed that chloride content of all samples was below its toxic limit, and cationic composition of water was dominated by calcium and magnesium ions.

For the determination of the quality of irrigation water, Flynn[42] considered pH, EC, calcium (Ca), magnesium (Mg), sodium (Na), chloride (Cl), Boron (B), sulfate (SO_4), and bicarbonate (HCO_3). Wastewater effluent is being used in drip irrigation with a good filtration system[43,44]

6.2.4 EMISSION UNIFORMITY, UNIFORMITY COEFFICIENT, AND COEFFICIENT OF VARIATION

Researchers have reported that application efficiency ranged from 92.67 to 93.12% for drip irrigation, and the uniformity coefficient (UC) ranged from 95.67 to 96.54%. Capra and Tamburino[7] computed the EU coefficient. Lamm[45] reported that for design and installation of micro irrigation system, equivalent uniformities were tabulated. The statistical uniformity was related to the field EU, by using the lower quartile method of calculating the distribution uniformity.

Soccol et al.[46] evaluated the performance of drip irrigation based on average EU, statistical uniformity, and coefficient of global variation, and these values were 74.51, 77.69, and 23.31%, respectively. The efficiency parameters were below expectations. Values of application efficiency, storage efficiency, deep percolation, deficit degree, and adequacy degree were 100, 47, 83, 52.17, and 0%, respectively.

Reinders[47] reported that field EU of all dripper types was deteriorated over time from 87 in the first evaluation to 82.4% in the fourth and last evaluation 1 year later. With regard to the statistical discharge coefficient, the drippers met in only 69% of the requirements.

El Gendy[48] stated that the direction of soil water movement was useful for defining the depth of sampling of active roots for water absorption and active rooting depth. These parameters clearly affected the water movement based on 100 and 75% of ET_c. Bakhsh et al.[49] reported that deficit irrigation has the potential to improve WUE for drip irrigation system. They compared effects of 15% (D_{15}) and 30% (D_{30}) deficit irrigation on WUE of cotton compared with no deficit (D_0) irrigation using drip irrigation system.

Aali and Liaghat[50] evaluated five types of emitters (with different nominal discharges with or without self-flushing system, and with or without pressure-compensating system) under three management practices: untreated well water, acidic treated water, and magnetic treated water. They studied effects of these parameters on chemical clogging, flow reduction rate, statistical UC, EU coefficient, and variation coefficient of emitter performance in the field.

Sah et al.[51] evaluated vegetable growth, hydraulic performance, crop water requirement, WUE, and cost economics for tomato and broccoli. The payback period was one season, and benefit-to-cost ratio was 1.59 to 5.31.

6.2.5 CLOGGING: CAUSES AND PREVENTION

Alam and Rogers[52] reported that all irrigation systems require proper maintenance. Subsurface drip irrigation systems are no exception. Clogging is major cause of failures in subsurface drip irrigation and other micro irrigation systems worldwide.

Bozkurt and Ozekici[53] evaluated the effects of different fertigation practices on clogging in in-line emitters using well water. Three different emitters and three different fertigation treatments with flushing and no flushing management groups were evaluated. Emitter discharge rates were tested at the beginning and at the end of every season to determine emitter flow variations, which depend on the degree of emitter clogging.

Ribeiro et al.[54] mentioned that many producers use trickle irrigation systems for flower production in the field and in protected environments. A frequent problem in drip irrigation is clogging of drippers, which is directly related to water quality and filtering efficiency.

Qingsong and Shuhuai[55] reported that emitter clogging will greatly affect the irrigation efficiency and the running cost of drip irrigation. If there is an effective method to predict emitter clogging, the loss will be reduced to a minimum. A solid–liquid two-phase turbulent model describing the flow within drip emitters was studied.

Liu and Huang[56] studied the emitter performance of three commonly used emitter types with the application of freshwater and treated sewage effluent. The three emitter types are the in-line-labyrinth types of emitters with turbulent flow (E_1) and laminar flow (E_2) and online pressure-compensating emitters (E_3). The qualities of freshwater and treated sewage effluent were measured, and the emitter performance was evaluated using the relative emitter discharge, the reduction of emitter discharge, the coefficient of variation, EU, Christiansen's uniformity coefficient, and the percentage of emitter clogging.

Yavuz et al[57] stated that emitter clogging affects the performance of drip irrigation. Emitter clogging is formed in a short time due to drip irrigation system running under an inadequate pressure or owing to water quality, and this negatively influences the uniformity of water distribution.

6.2.6 ACIDIFICATION TO PREVENT CLOGGING

Enciso and Porter[58] developed maintenance program for cleaning the filters, flushing the lines, adding chlorine, and injecting acids. If the preventive measures are taken, the need for major repairs such as replacing damaged parts often can be avoided to extend the life of the system.

Jain Irrigation[15] stated that precipitation of salts (calcium carbonate, magnesium carbonate, or ferric oxide) can cause either partial or complete blockage of the drip irrigation systems. Acid treatment is applied to prevent precipitation of such salts. Acid is also effective in cleaning systems, which are already blocked with precipitates of salts. Anonymous[3,15] stated that acid injection rate depends on the salt content in the water denoted by EC reading in a water test report. For EC above 0.5, acid injections are recommended. The frequency of acid injection for EC up to 1.0 can be 45 days; for EC 1.0–1.5, it should be 30 days; and for EC 1.5–2.5, it should be 15 days. Water of EC above 2 should not be used for drip irrigation.

Jain Irrigation[15] stated that the irrigation pumping plant and the chemical injection pump should be interlocked so that when irrigation pumping plant stops, chemical injection pump will also stop. It prevents chemical from the supply tank from filling irrigation lines if the irrigation pump stops. The system should be flushed regularly as determined by water quality, monitoring, and recording data. Start the flushing process from the pump onward. Make sure that the filters are clean and pressure is set correctly; systematically clean the mainline, submains, and laterals and flushing manifold.

Ribeiro et al.[54] indicated that the frequent problem in drip irrigation system is clogging of drippers, which is directly related to the water quality and the filtering system efficiency. They evaluated efficiency of nitric acid and sodium hypochlorite to unclog drippers that were clogged due to use of water with high algal content.

Aali and Liaghat[50] evaluated five types of emitters with different discharge rates, with or without self-flushing system, and with or without pressure-compensating systems. They used untreated well water, acidic treated water, and magnetic treated water in order to reduce chemical clogging.

Netafim USA[59] reported that to save money, concentrated and inexpensive technical acids should be used such as concentrated technical hydrochloric, nitric, or sulfuric acid. Phosphoric acid applied as fertilizer through the drip system might, under certain conditions, also act as a preventive measure against the formation of precipitates.

6.2.7 EFFECTS OF EMISSION UNIFORMITY ON CROP GROWTH PARAMETERS

Researchers have reported that increase in the plant density of capsicum from normal level can increase yield and thus can have more gross returns under greenhouse conditions. Rahman and El-Sheik[60] studied the effects of EU on growth parameter of three varieties (California Wonder, Yolo Wonder, and HW Yellow) of bell peppers and found that California Wonder was best suited for greenhouse cultivation as it gave 44,500 kg/ha under greenhouse compared with 16,400 kg/ha under field conditions. Similarly, among the three capsicum cultivars (Colombo, Galaxy, and Gideon) grown under plastic house, the cultivar Colombo gave the highest yield (4.17 kg/plant), fruit length (11.98 cm), fruit weight (230 g), diameter of fruit (7.75 cm), dry matter content (4.59%), and TSS (4.27%) compared with other varieties.

Seekar and Hochmuth[61] indicated that higher marketable yield of sweet pepper (4.62 kg/m^2) was obtained under plastic mulch compared with open (3.40 kg/m^2), and harvesting was earlier under plastic cover than in open field. Tunnel-covered plants resulted in higher yield (98.00 t/ha) compared with open field (68.00 t/ha) conditions.

Saen and Pathom[62] studied the effects of three pruning methods (no pruning, two-branch pruning, and four-branch pruning) on pepper yield and fruit quality. Pruning increased plant height, fruit weight, and fruit length. The four-branch pruning increased fruit weight by 13%.

Ashok[63] studied the effects of varying levels (50, 100, 150 ppm) and sources of N fertigation (ammonium nitrate, aqueous ammonia, nitric acid, and urea) on the flowering of cut rose cv. First Red under protected cultivation. Ammonium nitrate at 150 ppm gave the highest values of bud circumference (6.09 cm), flower diameter (7.33 cm), petal length (4.01 cm), petal breadth (3.84 cm), and flower yield (153/m^3) compared with other treatments.

Megharaja[64] observed significantly higher plant height (94.36 cm), number of branches (31.94), and total number of fruits (12.08) of capsicum cv. Indira under poly-house conditions compared with plants grown under open field conditions (45.33, 14.25, and 5.43 cm, respectively). Higher values of fruit length, fruit breadth, fruit weight, and fruit volume (8.54 cm, 6.76 cm, 120.06 g and 255.97 cm^3, respectively) were recorded under greenhouse conditions compared with those procured from open field (5.67 cm, 4.40 cm, 56.17 g, and 114.42 cm³, respectively).

Barbosa and Santose[65] evaluated flower yield and quality of rose cv. Sonia and Red Success as affected by different potassium (K) rates through drip irrigation. The K treatment was 0, 30, 60, and 90 g/m²/year applied to the plants. Maximum yield of rose stems with superior quality (>69 cm in length) was obtained with a K application rate of 49.76 g/m²/year. Sonia yielded a higher number of normal straight stem of 50.36 g/m²/year. The increase in K application in Red Success produced a reduction in the infection of diseases.

Muhammad[66] studied eight cultivars of gladiolus to study the influence of potassium levels (0, 100, and 200 kg/ha). All growth parameter were significantly affected by different potassium levels. Potassium levels significantly affected days to spike emergence and opening of first florets. A spike emergence was earlier at 100 kg/ha.

6.3   MATERIALS AND METHODS

During January 2011 to July 2011, a field experiment on drip irrigation system was installed in poly-house, at Hi-Tech Floriculture Project, Fruit Research Station, Aurangabad. Aurangabad district is situated in the Marathwada region of Maharashtra state at 19°N latitude and 20°E longitude. The average rainfall at the site was 734 mm. The temperature ranged from 5.6 to 45.9°C.

The experiments were planned in two poly-houses: one poly-house with online-type pc emitters (orifice) and other poly-house with in-line-type non-pc emitters (long path). Each poly-house was of 20 R each with irrigation hydraulic valve that was for one sector. There were six treatments with six plots in each poly-house. In one poly-house, pot culture in cocopeat (substrate) cultivation was used with Netafim online pressure-compensating dripper. Other poly-house was provided with soil cultivation (Red soil mix) with Jain in-line non-pressure-compensating dripper. Drip irrigation system consisted of four valves and fertigation unit. The experimental layouts are shown in Figs. 6.1 and 6.2. Five types of acids were evaluated (Figs. 6.3 and 6.4). Treatments consisted of the following:

Main Treatments	
Online-type pressure-compensating emitters (orifice)	In-line-type non-pressure-compensating emitters (long path)
Sub-treatments	

OT_1 – DS-99	LT_1 – DS-99
OT_2 – Sulfuric acid (H_2SO_4) – 75%	LT_2 – Sulfuric acid (H_2SO_4) – 75%
OT_3 – Nitric acid (HNO_3) – 70%	LT_3 – Nitric acid (HNO_3) – 70%
OT_4 – Hydrochloric acid (HCl) – 37%	LT_4 – Hydrochloric acid (HCl) – 37%
OT_5 – Phosphoric acid (H_3PO_4) – 85%	LT_5 – Phosphoric acid (H_3PO_4) – 85%
OT_6 – Flushing interval 15 days	LT_6 – Flushing interval 15 days

The observations consisted of the following:

First observation at 0 days	Average EU for before and after each treatment
Second observation at 15 days	Average EU for before and after each treatment
Third observation at 30 days	Average EU for before and after each treatment

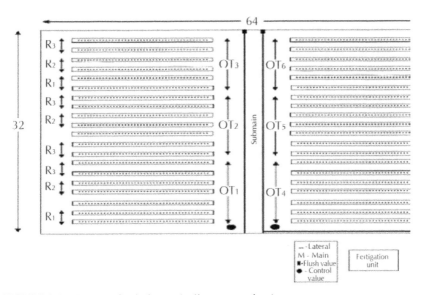

FIGURE 6.1   Layout of poly-house (online-type emitter).

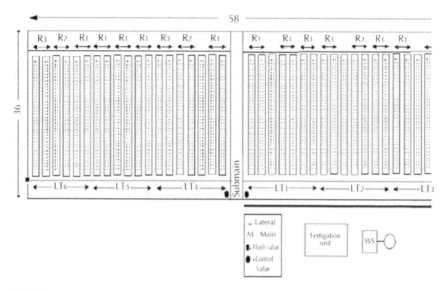

FIGURE 6.2   Layout of poly-house (in-line-type emitter).

FIGURE 6.3   DS-99 acid and phosphoric acid.

FIGURE 6.4   Hydrochloric acid, sulfuric acid, and nitric acid used for treatments.

6.3.1   DESIGN SPECIFICATIONS

Tables 6.2 and 6.3 show details of plots in poly-house of in-line-type emitter (non-pc) and poly-house of online-type emitter (pc). Design specifications are shown below:

Type of Poly-House	Partially Controlled Naturally
	Ventilated poly-house
Experimental design	Random block design (RBD)
Number of replications	3
Number of poly-houses	2
Size of poly-house with online-type emitter	2048 m² (64 m × 32 m)
Size of poly-house with in-line-type emitter	2088 m² (58 m × 36 m)
Direction	North–South
Sector of two poly-houses	4
Plots per sector	3
Total plots in two poly-houses	12
Size of each plot	3.3 Area (R)
Number of valves	4
Number of valves per sector	1
Fertigation unit	Make Talgil, from Israel.
Water source	Tube well

TABLE 6.2　Details of Plot in Each Poly-House of Online-Type Emitters (pc)

Particulars	Treatments					
	OT_1	OT_2	OT_3	OT_4	OT_5	OT_6
Area (R)	3.3	3.3	3.3	3.3	3.3	3.3
No. of lines	7	7	6	7	7	6
No. of laterals	7	7	6	7	7	6
No. of drippers per lateral	96	96	96	91	91	91
Discharge of dripper (lph)	8	8	8	8	8	8
Arrow dripper with four-way mani-fold (lph)	2	2	2	2	2	2
Volume of water (lph)	5376	5376	4608	5096	5096	4368
Volume of water (L/5 min)	448	448	384	424.6	424.6	364
Acid required for 1 L of water (mL)	0.8	0.8	0.8	0.8	1	–
Acid required for volume of water (mL/5 min)	358.4	358.4	307.2	339.6	424.6	–
Interval of treatments (days)	15	15	15	15	15	15

TABLE 6.3　Details of Plot in Each Poly-House of In-line-Type Emitter (Non-pc)

Particulars	Treatments					
	LT_1	LT_2	LT_3	LT_4	LT_5	LT_6
Area (R)	3.3	3.3	3.3	3.3	3.3	3.3
No. of beds	7	7	7	7	6	6
No. of laterals	14	14	14	14	12	12
No. of drippers per lateral	158	158	158	158	158	158
Discharge of dripper (lph)	1.3	1.3	1.3	1.3	1.3	1.3
Volume of water (lph)	2875.6	2875.6	2875.6	2875.6	2464.8	2464.8
Volume of water (L/5 min)	239.6	239.6	239.6	239.6	205.4	205.4
Acid required for 1 L of water (mL)	0.8	0.8	0.8	0.8	1.0	–
Acid required for volume of water (mL/5 min)	191.68	191.68	191.68	191.68	205.4	–
Interval of treatments (days)	15	15	15	15	15	15

6.3.2 DRIP IRRIGATION SYSTEM

In this chapter, the drip irrigation system consisted of water source, pump, control valves, sand filter, pressure gauges, fertigation unit, mainline, submain line, lateral, emitters, and other accessories (tee, elbow, coupling, flush valve, end plug, etc.).

The existing water source on farm is a bore well that is located adjacent to the experimental plot. The 10 HP submersible pump was used for pumping the water from the sump well. Control valve consisted of bypass arrangement, non-return valve (NRV), pressure release valve (PRV), and gate valve. NRV and PRV were used to control the flow and pressure in the drip irrigation system. Bypass was used to divert the water in excess of the design discharge to the well by opening the bypass valve. It was provided before the filter. Gate valve was used to create the pressure difference for the fertigation units. Sand filter was used to remove organic and inorganic contaminates from the source of bore well water. Pressure gauge was used to check and measure the water pressure developed in pipe lines. In drip irrigation system, 1.5 kg/cm² of operating pressure was provided at head unit. Fertigation unit make Talgil (imported from Israel) was used (Fig. 6.5). It consisted of 200-L tank, 7.5 HP motor, venturi meter, pH sensor, and EC sensor. Water meter measured the quantity of water flowing through the system.

FIGURE 6.5 Fertigation unit.

Mainline carried irrigation water from the head unit to submain. The mainline was polyvinyl chloride (PVC) pipe with 75 mm diameter and 4 kg/cm² pressure class. Submain line carried irrigation water from mainline to laterals. PVC pipe of

63 mm diameter and 4 kg/cm^2 pressure class was used. Lateral carried water from submain to each crop. The low-density polyethylene pipes of 16 mm diameter were used as lateral. The lateral spacing in poly-house with substrate cultivation was 1.6 m, and poly-house with soil cultivation was 1.2 m. Dripper is the water-emitting device to convey water from lateral to the root zone of crop. Online- and in-line-type drippers were used. In this study, authors used online-type pressure-compensating dripper having a discharge of 8 lph with four-way manifold spaced at 30 cm, the arrow dripper connected to the four-way manifold of main dripper with a micro tube of 4 mm diameter, and in-line-type non-pressure-compensating dripper having a discharge of 1.3 lph spaced at 20 cm.

Accessories (PVC coupling, bends, tees, elbows, end cap, and flush valve) were used to make necessary connections in mainline and submain lines. End plugs were used at the end of mainline and submain line to plug the water flow.

6.3.3 PERFORMANCE EVALUATION OF DRIP IRRIGATION SYSTEM

In order to evaluate the performance of drip irrigation system in this chapter, EU was recorded. The observations for discharges through the emitters were recorded by using the following standard procedure:

- For EU, six laterals were selected from 3.3-R area at first two lines, middle two lines, and last two lines. Random emitters (two from each line) were selected.
- Discharge of the selected emitter for 15 min was collected in the measuring cylinder after starting the pulse.
- Observed values were arranged in ascending or descending order.
- The average of the lowest one-fourth value of descending or ascending order data was calculated. Then, the average of all the values of descending or ascending order data was determined.

a. Emission uniformity
The EU was determined by using the following equation[7]:

$$EU = 100 \times [q_n/q_a] \tag{1}$$

where EU is the field emission uniformity (%), q_n is the average of the lowest one-fourth of the field data on emitter discharge (lph), and q_a is the average of all the field data emitter discharge (lph). Procedure to measure EU is shown in Figs. 6.6–6.9.

FIGURE 6.6   Selected orifice-type emitter (online pc) for measuring emission uniformity.

FIGURE 6.7   Selected in-line-type emitter (non-PC) for measuring emission uniformity.

FIGURE 6.8   Measurement of discharge for emission uniformity on orifice-type emitter (online pc).

FIGURE 6.9   Measurement of discharge for emission uniformity on in-line-type emitter (non-pc).

b. **Uniformity coefficient**

The UC (%) was determined as follows[67]:

$$UC = 100 \times [1 - (S_q/q_a)]\ \ \ \ \ \ \ \ \ \ (2)$$

where UC is the uniformity coefficient (%), S_q is the standard deviation of emitter discharge (lph), and q_a is the average emitter discharge (lph).

c. Coefficient of Variation

The coefficient of variation (C_v, %) is determined as follows[68]:

$$C_v = 100 \times [S_q/q_a] \tag{3}$$

where C_v is the coefficient of variation (%), Sq is the standard deviation of emitter discharge (lph), and q_a is the average emitter discharge (lph). Classification of performance of drip irrigation system is shown in Table 6.4.

TABLE 6.4 Classification of Performance of Drip Irrigation System

Particulars	Performance Evaluation (%)				
	Excellent	Good	Marginal	Poor	Unacceptable
Emission uniformity	94–100	81–87	68–75	56–62	<50
Uniformity coefficient	95–100	85–90	75–80	65–70	<60
Coefficient of variation	<5	5–7	7–11	11–15	>15

Source: From ASAE-EP458.1.[69]

6.3.4 IMPLEMENTATION OF ACID TREATMENT

The acid required for the known volume of water sample was determined by the following procedure:

1   For acid treatment, water sample of 1 L was collected from the existing water source of the project.
2.   Simple titration method was followed by adding acid drop by drop in this water sample. At the time of titration, glass rod is used frequently for stirring, and pH of the solution was determined.
3.   The quantity of acid required to maintain a pH value of 4.00 was calculated. The pH meter was used to determine the pH value.
4.   To implement the acid treatment, fertigation unit was used. The acid was mixed with water, and the diluted solution was discharged at the other end.
5.   The unit was kept 24 h un-operated, and the action of acid on clogged laterals was observed. Generally, acid action will be effective after 6–8 h of discharge.
6.   After 24 h, the unit was operated to flush the submains and laterals so that the remaining residues of the salts will be driven out of the system, and the EU can be effectively observed (Fig. 6.10).

FIGURE 6.10   Flushing of laterals after acid treatment.

6.3.5   ACID INJECTION RATE

The acid injection rate can be determined by the following equation:

$$Q_a = [3.6 \times Q \times A]/[V] \tag{4}$$

where Q_a is the acid injection rate (lph), Q is the system flow rate (lph), A is the acid quantity (mL) to achieve the required pH in a water sample, and V is the Volume of test sample (liters).

6.3.6   BIOMETRIC OBSERVATIONS

It was found that the emitters were unclogged after acidification. To study the crop response to acidification, the biometric observations of Dutch rose flower in sub-strate and capsicum (bell pepper) in soil were recorded.

6.3.6.1   BIOMETRIC OBSERVATIONS: DUTCH ROSE FLOWER

The crop growth parameters (number of sprouts, stalk height, and stalk girth) were recorded approximately at 60-day interval after planting until harvest with five randomly selected plants in each replication from each plot. The observations at harvest included bud length, bud diameter, and the number of flowers/m². During 2011, the cultural operations were conducted as follows:

February 10: Field layout   April 6–9 Sprouting

February 16 Transplanting   June 26–July 10 Harvesting, six

March 23 Bending

Under substrate cultivation, irrigation and fertigation are a single activity. There-fore, pH and EC values of the irrigation water were fixed in the range of 5.5–6 and 1.2–1.4, respectively. There were four to six irrigation pulses of 3-min duration per day, monitored on drip drain ratio strictly to 33%. On the basis of water analysis, fertilizer dose per 1000 L of base solution was formulated as shown below:

Tank A	Tank B	Tank C
1. Calcium nitrate 57 kg 2. Ferrous (EDDHA) 2.8 kg	1. Potassium nitrate 30 kg 2. Potassium sulfate 22 Kg 3. Monoammonium phosphate 14 kg 4. Magnesium sulfate 85 g 5. Zinc sulfate 115 g 6. Borax 286 g 7. Copper sulfate 19 g 8. Ammonium molybdate 12 g	Contains acid solution to control pH of irrigation to the desired level

During the crop growth, the biometric observations were recorded on regular intervals manually. At 15 days after bending operation, the number of sprouts for selected plants in each replication was observed, and initiated sprouts were recorded (Fig. 6.11). After the commencement of sprouting, the length of sprout was recorded by using a ruler at an interval of 10 days, until the harvesting (Fig. 6.12). Stalk girth of a flower is the diameter of the growing sprout, and it was recorded with a vernier caliper at an interval of 10 days until harvesting (Fig. 6.13). After maturity of stalk buds, bud length from the base of the flower to tip of the flower was recorded with a vernier caliper (Fig. 6.14). After attaining full distance between two horizontal ends, bud diameter of the flower was recorded. The number of flowers/m^2 at each harvest was recorded.

FIGURE 6.11   Sprouts of Dutch roses after bending.

FIGURE 6.12   Measurement of stalk length.

FIGURE 6.13   Measurement of stalk girth.

FIGURE 6.14   Measurement of bud length.

6.3.6.2   BIOMETRIC OBSERVATIONS OF CAPSICUM

The crop growth parameters were recorded from five randomly selected plants in each replication from each plot (Figs. 6.15–6.19). Plant height was recorded at 30-day interval after transplanting until last harvest. Plant spread (EW/NS) and plant girth were recorded approximately at 60-day interval after transplanting until harvest. These plants were properly labeled, and growth parameters were monitored. The observations at harvest included fruit length, fruit breadth, number of fruit, and fruit yield/m^2. The cultural operations were scheduled as follows:

Operations	Frequency	Date of Operation
Sowing of seed	1	February 13, 2011
Field layout	–	February 10, 2011
Transplanting	1	March 31, 2011
Harvesting	3	June 27, 2011, to July 10, 2011

FIGURE 6.15   Measurement of capsicum plant height.

FIGURE 6.16   Measurement of capsicum plant spreads.

FIGURE 6.17   Measurement of capsicum plant girth.

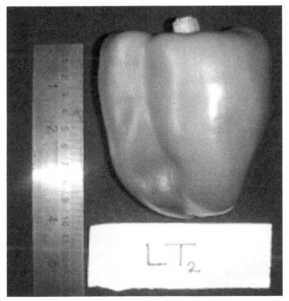

FIGURE 6.18   Fruit length.

FIGURE 6.19   Fruit weight.

The fertigation (kg/ha/day) as per recommendations for bell pepper was done in combination with irrigation to obtain higher yield and better fruit by increasing application efficiency. Fertilizer dose for capsicum per 1000 m² was as follows:

Fertilizer Recommended Dose

19:19:19 1 kg

00:52:34 200 g

00:00:50 375 g

After transplanting bell pepper, plant height (cm) was measured with a ruler at 30 days after transplanting (DAP) interval. Plant spread (cm) in east–west and north–south directions was recorded with a ruler at 60 DAP interval. After 60 DAP, plant girth (cm) was measured with a vernier caliper at fourth night interval. After 90 DAP until last harvest, fruits were harvested, and numbers of fruits harvested per plant were counted. After harvesting, the fruit weight (kg per plant) was recorded with a digital weighing scale. After harvesting, fruit length (the distance between two terminal ends) and breadth (distance between two broader ends) were measured with a vernier caliper. Fruit shape index was calculated as follows:

$$\text{Shape index} = [\text{length} \times \text{breadth}]/2 \qquad (5)$$

After harvesting, fruit weight per plant was observed with a digital weighing balance in kg, and the yield (kg/m^2) was determined.

6.3.6.3   STATISTICAL ANALYSIS

All data were analyzed using MAU-STAT software. The appropriate standard error (SE) and the critical difference (CD) were calculated at $P = 5\%$.

6.4   RESULTS AND DISCUSSION

Partially clogged drip irrigation field was operated to evaluate the effects of acidification on growth performance of Dutch rose flower and bell pepper, at the Hi-Tech Floriculture Project of Fruit Research Station in Aurangabad, India. The water quality analysis was carried out in the laboratory of the Department of Soil Science and Agriculture Chemistry, College of Agriculture, MKV, Parbhani, India.

6.4.1. CHEMICAL ANALYSIS OF WATER

The source of water was tube well. The water was analyzed for the dissolved salts, EC, pH, and SAR. These values are reported in Table 6.5.

TABLE 6.5   Chemical Analysis of Water

Particulars	Units	Safe Limits	Observed Value
Bicarbonate	mEq/L	0–1.50	8.4
Calcium	mEq/L	0.0 – 10	44.0
Carbonate	mEq/L	0–1.50	1.2
Chloride	mEq/L	0–2.00	27.0
Electrical conductivity, EC	ms/cm	0–0.25	0.645
Magnesium	mEq/L	0–1.25	56.0
pH	–	6.5–7.5	7.2
Residual sodium carbonate	–	<1.25	<1.25
Sodium absorption ratio (SAR)	–	<10	<10

The chemical analysis of water source revealed that the water contains mostly neutral salts such as NaCl, CaCl$_2$, and MgCl$_2$. The carbonate content of water was within safe limits, and the bicarbonate content of the water was moderately high.

The water was of Mg–Ca cationic and Cl–HCO$_3$ anionic dominant type. The SAR of water was within the safe limits of <10.[70] The pH content of water was 7.2 thus indicating slightly alkaline nature of water. The EC of water was 0.645 ms/cm. Continuous use of such water may clog the drippers due to precipitation of CaCO$_3$ and MgSO$_4$.

English[71] has reported that water supply containing salts of calcium, iron, and sulfur may precipitate and can cause clogging.

6.4.2   QUANTITY OF ACIDS REQUIRED FOR 1 L OF WATER SAMPLE

Each acid at different concentration was titrated against 1 L of water sample to lower down its pH from 7.2 to 4. The quantity of acid required to lower down pH to 4.00 of 1 L of water is given in Table 6.6.

TABLE 6.6   Quantity of Acid Required for 1 L of Water Sample

Type of Acid	Acid Concentration, %	pH	Quantity of Acid for 1 L of Water Sample, mL
DS	99	4	0.8
HNO$_3$	70	4	1
H$_2$SO$_4$	75	4	0.8
HCl	37	4	0.8
H$_3$PO$_4$	85	4	0.8

Lowering of pH of water prevents precipitation of salts. Similar observations were reported by Darbie.[72]

6.4.3   PERFORMANCE OF DRIP IRRIGATION SYSTEM

Two selected drip irrigation sets from two poly-houses (online pressure-compensating-type emitters (pc) and in-line non-compensating-type emitter (non-pc)) were evaluated by collecting the emitter flow for 1 month (January 1, 2011, to February 2, 2011). From the recorded observations, evaluation of performance parameters included EU, UC, and coefficient of variations.

6.4.3.1 ONLINE-TYPE PC EMITTER

6.4.3.1.1 AVERAGE EMISSION UNIFORMITY BEFORE AND AFTER THE FIRST APPLICATION OF ACID

The average EU before and after the first application of acid is presented in Table 6.7 and Fig. 6.20.

TABLE 6.7 Average Emission Uniformity before and after the First Application of Acid

Treatments	Emission Uniformity, EU, %	
	Before	**After**
OT_1	74.98	82.21
OT_2	74.30	87.71
OT_3	74.50	84.12
OT_4	73.57	85.58
OT_5	73.74	81.60
OT_6	72.83	75.21
SE ±	0.56	0.45
CD at $P = 5\%$	NS	1.42

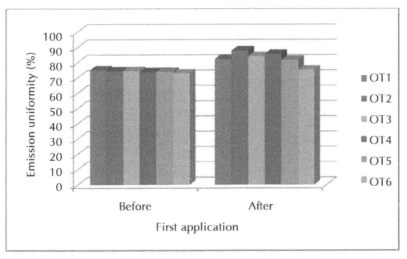

FIGURE 6.20 Emission uniformity (%) for online-type emitter, after the first application of acid.

Table 6.7 indicated that average EU before the first application for online-type emitter (pc) ranged from 72.83 to 74.98%. This limit of EU was poor for drip irrigation system. Consequently, the hazard of chemical clogging under saline water usage can be reduced by acidification, and high performance can be promised. Results indicate that the system was under clogging conditions due to deposition of salts and the salts (chlorides, calcium, magnesium, and sodium) present in water. The carbonate and bicarbonate salts were also present. Clogging was also caused by the type of emitter, flow regime, and energy dissipation pattern in the emitter. These results were nonsignificant for drip irrigation system.

After acid and flushing treatments, there was a slight change in the EU of drip irrigation system. It was observed from Table 6.7 that average EU after the first application of acid treatment for online-type emitter (pc) ranged from 81.60 to 87.71%. This range of EU was good for drip irrigation system, due to dissolution of salts in acids. From Table 6.7, average EU after the first application of flushing treatment (control) for online-type emitter (pc) was found to be 75.21%. Similar observations were observed by Sah et al.,[51] and category-wise performance evaluation was presented. This limit of EU is considered marginal for drip irrigation system. Results were significant for drip irrigation system. The system was kept running under continuous conditions.

6.4.3.1.2   AVERAGE UNIFORMITY COEFFICIENT BEFORE AND AFTER THE FIRST APPLICATION OF ACID

It can be observed from Table 6.8 that average UC before the first application of acid for online-type emitter (pc) was in the range of 79.89 to 80.84%. This limit of UC was marginal for drip irrigation system.

TABLE 6.8 Average Uniformity Coefficient before and after the First Application of Acid

Treatments	Uniformity Coefficient, UC, %	
	Before	**After**
	First Application	
OT_1	80.84	87.69
OT_2	80.45	90.91
OT_3	80.70	89.09
OT_4	80.28	89.64
OT_5	80.54	87.48
OT_6	79.89	81.30

After acid and flushing treatments, there was a slight change in the UC of drip system. It was observed from Table 6.8 that average UC after the first application of acid treatments for online-type emitter (pc) was in the range of 87.4–90.91%. This limit of UC was good for drip irrigation system. Average UC after the first application of flushing treatment (control) for online-type emitter (pc) was 81.30%. Similar observations were recorded by other investigators[3] who presented category-wise performance evaluation for UC. This limit of UC was marginal for drip irrigation system.

6.4.3.1.3 AVERAGE COEFFICIENT OF VARIATION BEFORE AND AFTER THE FIRST APPLICATION OF ACID

It can be observed from Table 6.9 that average coefficient of variation before the first application of acid for online-type emitter (pc) was in the range of 20.11–19.16%. This limit of coefficient of variation is unacceptable for drip irrigation system.

TABLE 6.9   Average Coefficient of Variation before and after the First Application of Acid

Treatments	First Application of CV (%)	
	Before	**After**
	First Application of Acid	
OT_1	19.16	12.31
OT_2	19.55	9.09
OT_3	19.30	10.91
OT_4	19.72	10.36
OT_5	19.46	12.52
OT_6	20.11	18.70

After acid treatments and flushing treatment, there was slight change in the coefficient of variation of drip system. It was observed from Table 6.9 that average coefficient of variation after the first application of acid treatments for online-type emitter (pc) were 9.09, 10.91, and 10.36% for OT_1, OT_2, and OT_3, respectively. Table 3.3 shows that this limit of coefficient of variation was marginal for drip irrigation system. Also treatments OT_1 and OT_5 had the coefficient of variation 12.31 and 12.52, respectively. This limit of coefficient of variation was poor for drip irrigation system. From Table 6.9, average coefficient of variation after the first application of flushing treatment (control) for online-type emitter (pc) was found 18.70%. This limit of UC was unacceptable for drip irrigation system. Similar observations were recorded by Reinders.[47] The average discharge variation CVq was a poor 8.2% with a variation of 2.7% up to 22.2% for the individual drip lines. The category-wise performance evaluation for coefficient of variation was presented.

6.4.3.1.4 AVERAGE EMISSION UNIFORMITY BEFORE AND AFTER THE SECOND APPLICATION OF ACID

The average EU before and after the second application of acid is presented in Table 6.10 and Fig. 6.21.

TABLE 6.10 Average Emission Uniformity, before and after the Second Application of Acid

Treatments	EU (%)	
	Before	**After**
	Second Application of Acid	
OT_1	82.09	92.92
OT_2	87.45	98.11
OT_3	83.85	94.22
OT_4	85.23	95.70
OT_5	81.28	92.18
OT_6	74.76	75.33
SE ±	0.424	0.46
CD at 5%	1.335	1.448

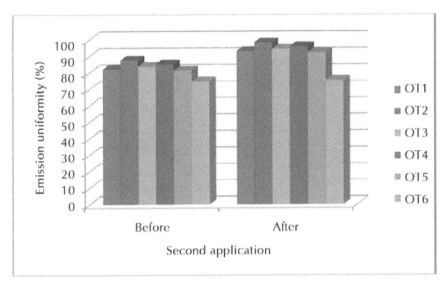

FIGURE 6.21   Emission uniformity (%) for online-type emitter, for the second application of acid.

It can be observed from Table 6.10 that average EU before the second application of acid for online-type emitter (pc) was in the range of 81.28–74.76%. This limit of EU was marginal for drip irrigation system due to continuous flow of water inside the laterals and emitters for a period of 15 days. These results were significant for drip irrigation system.

After acid and flushing treatments, there was a slight change in the EU of drip system. The highest EU was 98.11% for OT_2 treated with sulfuric acid, thus indicating dominant effect of sulfuric acid compared with all other treatments. This limit of EU was excellent for drip irrigation system, due to dissolution of salts in acids. From Table 6.10, it can be observed that average EU after the second application of flushing treatment (control) for online-type emitter (pc) was 75.33%. This limit of EU was marginal for drip irrigation system. These results were significant for drip irrigation system. Similar observations were recorded by other investigators[3] for EU.

6.4.3.1.5 AVERAGE UNIFORMITY COEFFICIENT BEFORE AND AFTER THE SECOND APPLICATION OF ACID

Table 6.11 reveals that the highest UC was 90.82% for OT_2 treated with sulfuric acid, which shows that the effect of sulfuric acid treatment was dominant compared with all other treatments. This limit of UC was good for drip irrigation system.

TABLE 6.11   Average Uniformity Coefficient before and after the Second Application of Acid

Treatments	UC (%)	
	Before	**After**
	Second Application of Acid	
OT_1	87.63	95.26
OT_2	90.82	98.31
OT_3	88.96	96.07
OT_4	89.49	96.76
OT_5	87.34	94.91
OT_6	81.13	81.41

After acid and flushing treatments, there was a slight change in the UC of drip system. It can be observed in Table 6.11 that average UC after the second application of acid treatments for online-type emitter (pc) ranged from 94.91 to 98.31%. This limit of UC was excellent for drip irrigation system. From Table 6.11, it can be observed that average UC after the second application of flushing treatment (control) for online-type emitter (pc) was 81.41%. This limit of UC was marginal for drip irrigation system. The present results were in agreement with other investigators.[3]

6.4.3.1.6 AVERAGE COEFFICIENT OF VARIATION BEFORE AND AFTER THE SECOND APPLICATION OF ACID

Table 6.12 reveals that the highest coefficient of variation was 9.18% for OT_2 treated with sulfuric acid. Therefore, the effect of sulfuric acid treatment was dominant compared with all other treatments. This limit of coefficient of variation was marginal for drip irrigation system. The lowest coefficient of variation was 18.87% for OT_6 treated with flushing (control). This limit of coefficient of variation was unacceptable for drip irrigation system.

TABLE 6.12 Average Coefficient of Variation before and after the Second Application of Acid

	CV (%)	
	Before	**After**
Treatments	**Second Application of Acid**	
OT_1	12.37	4.74
OT_2	9.18	1.69
OT_3	11.04	3.93
OT_4	10.51	3.24
OT_5	12.66	5.09
OT_6	18.87	18.59

After acid and flushing treatments, there was a slight change in the coefficient of variation of drip system. Results revealed that the highest coefficient of variation was 1.69% for OT_2 treated with sulfuric acid, thus indicating dominance of this acid compared with all other treatments. This limit of coefficient of variation was excellent for drip irrigation system. The lowest coefficient of variation was 18.59% for OT_6 treated with flushing (control). This limit of coefficient of variation was unacceptable for drip irrigation system. Similar observations were recorded by other investigators[3] for coefficient of variation.

Online-type pressure-compensating emitter (orifice) had high discharge of 8 lph and turbulent flow regime. Therefore, the system was less susceptible to clogging although the drip irrigation system was 7-year old. Also it gave good response to acid treatments, and emitters were reclaimed after the second application of acid to the highest level. Flushing treatment (control) gave less response compared with acid treatment.

6.4.3.2 IN-LINE-TYPE EMITTER – NON-PC

6.4.3.2.1 AVERAGE EMISSION UNIFORMITY BEFORE AND AFTER THE FIRST APPLICATION OF ACID

The data on average EU before and after the first application of acid are presented in Table 6.13 and Fig. 6.22.

TABLE 6.13   Average Emission Uniformity before and after the First Application of Acid

	EU (%)	
	Before	**After**
Treatments	**First Application of Acid**	
LT$_1$	34.69	53.24
LT$_2$	34.29	59.82
LT$_3$	34.31	55.57
LT$_4$	35.95	56.65
LT$_5$	34.20	51.96
LT$_6$	34.69	37.74
SE ±	0.652	0.505
CD at 5%	NS	1.591

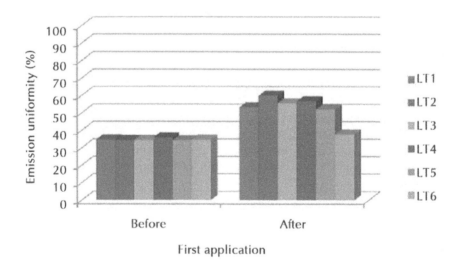

FIGURE 6.22   Emission uniformity (%) for in-line-type emitter before and after the first application of acid.

It can be observed from Table 6.13 that average EU before the first application of acid for in-line-type emitter (non-pc) ranged from 34.20 to 35.95%. This limit of EU was unacceptable for drip irrigation system. This indicates that the system was under severe clogging conditions, due to deposition of salts and the salts present in water (chlorides, calcium, magnesium, and sodium). The carbonate and bicarbonate salts were also present. Clogging was also due to the type of emitter, flow regime, and energy dissipation pattern in the emitter. These results were nonsignificant for drip irrigation system.

After acid and flushing treatments, there was a slight change in the EU of drip system. It was observed from Table 6.13 that average EU after the first application of acid treatment for in-line-type emitter (non-pc) ranged from 51.96 to 59.82%. This limit of EU was poor for drip irrigation system, due to the presence of salts in water. Table 6.13 reveals that average EU after the first application of acid in LT_6 treatment (control) for in-line-type emitter (non-pc) was 37.74%. This limit of EU was unacceptable for drip irrigation system. These results were significant for drip irrigation system. The system was kept in running condition continuously. Similar observations were recorded by other investigators.[3]

6.4.3.2.2   *AVERAGE UNIFORMITY COEFFICIENT BEFORE AND AFTER THE FIRST APPLICATION OF ACID*

Table 6.14 shows that average UC before the first application of acid for in-line-type emitter (non-pc) was in the range of 54.25–56.02%. This limit of UC was poor for drip irrigation system.

After acid and flushing treatments, there was a slight change in the UC of drip system. It can be observed from Table 6.14 that average UC after the first application of acid treatment for in-line-type emitter (non-pc) was in the range of 61.60–70.48%. This limit of UC was marginal for drip irrigation system. Table 6.14 shows that average UC after the first application of flushing treatment (control) for in-line-type emitter (non-pc) was 56.52%. This limit of UC was unacceptable for drip irrigation system. The present results were in accordance with other investigators[3] for UC.

TABLE 6.14   Average Uniformity Coefficient before and after the First Application of Acid

Treatments	UC (%)	
	Before	**After**
	First Application of Acid	
LT_1	52.32..	62.49
LT_2	54.34	70.48
LT_3	54.25	68.72
LT_4	55.85	68.70
LT_5	56.02	61.60
LT_6	55.20	56.52

6.4.3.2.3 AVERAGE COEFFICIENT OF VARIATION BEFORE AND AFTER THE FIRST APPLICATION OF ACID

Table 6.15 shows that average coefficient of variation before the first application of acid for in-line-type emitter (non-pc) was in the range of 43.98–47.68%. This limit of coefficient of variation was unacceptable for drip irrigation system.

After acid and flushing treatments, there was a slight change in the coefficient of variation of drip system. It was observed from Table 6.15 that average coefficient of variation after the first application of acid treatments for in-line-type emitter (non-pc) was in the range of 29.52–43.48. This limit of UC was unacceptable for drip irrigation system.

TABLE 6.15   Average Coefficient of Variation before and after the First Application

| Treatments | CV (%) | |
| | Before | After |
	First Application of Acid	
LT_1	47.68	37.51
LT_2	45.66	29.52
LT_3	45.75	31.28
LT_4	44.15	31.30
LT_5	43.98	38.40
LT_6	44.80	43.48

6.4.3.2.4 AVERAGE EMISSION UNIFORMITY BEFORE AND AFTER THE SECOND APPLICATION OF ACID

The data on average EU before and after the second application of acid are presented in Table 6.16 and Fig. 6.23.

TABLE 6.16   Average Emission Uniformity before and after the Second Application of Acid

Treatments	EU (%)	
	Before	**After**
	Second Application of Acid	
LT$_1$	53.08	71.93
LT$_2$	59.82	81.05
LT$_3$	54.90	74.29
LT$_4$	55.56	77.00
LT$_5$	51.42	71.15
LT$_6$	37.21	38.98
SE ±	0.472	0.709
CD at 5%	1.485	2.231

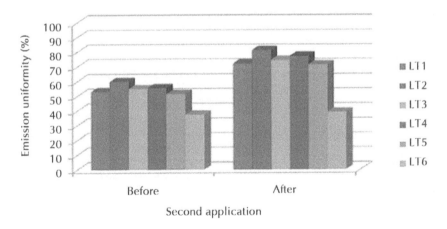

FIGURE 6.23   Emission uniformity (%) for in-line-type emitter, before and after the second application of acid.

It can be observed in Table 6.16 that average EU after the second application of acid treatments for in-line-type emitter (non-pc) was in the range of 51.42–59.82%. This limit of EU was poor for drip irrigation system, due to the presence of salts in water. Table 6.16 shows that average EU after the second application of acid in LT_6 treatment (control) for in-line-type emitter (non-pc) was 37.21%. This limit of EU was unacceptable for drip irrigation system.

After acid and flushing treatments, there was a slight change in the EU of drip system. The highest EU was 81.05% for LT_2 treated with sulfuric acid, thus indicating dominance of sulfuric acid treatment compared with all other treatments. This limit of EU is marginal for drip irrigation system, due to the dissolution of salts in acids. Table 6.16 shows that average EU after the second application of flushing treatment (control) for in-line-type emitter (non-pc) was 38.98%. This limit of EU was unacceptable for drip irrigation system. These results were significant for drip irrigation system.

6.4.3.2.5 AVERAGE OF UNIFORMITY COEFFICIENT BEFORE AND AFTER THE SECOND APPLICATION OF ACID

Table 6.17 shows that the highest UC was 70.48% was for LT_2 treated with sulfuric acid, thus indicating superiority sulfuric acid treatment compared with all other treatments. This limit of UC was marginal for drip irrigation system.

After acid and flushing treatments, there was a slight change in the UC of drip system. It was observed from Table 6.17 that average UC after the second application of acid treatments for in-line-type emitter (non-pc) was in the range of 81.31 to 84.90%. This limit of UC was good for drip irrigation system. As shown in Table 6.17, average UC after the second application of flushing treatment (control) for in-line-type emitter (non-pc) was 56.22%. This limit of UC was unacceptable for drip irrigation system.

TABLE 6.17 Average Uniformity Coefficient before and after the Second Application of Acid

Treatments	UC (%)	
	Before	**After**
	Second Application of Acid	
LT_1	62.78	81.31
LT_2	70.48	84.90
LT_3	68.01	82.65
LT_4	68.42	82.50
LT_5	61.32	81.50
LT_6	56.02	56.22

6.4.3.2.6   AVERAGE COEFFICIENT OF VARIATION BEFORE AND AFTER THE SECOND APPLICATION OF ACID

It can be observed from Table 6.18 that average coefficient of variation before the second application for in-line-type emitter (non-pc) was in the range of 29.52–43.98%. This limit of coefficient of variation was unacceptable for drip irrigation system.

After acid and flushing treatments, there was a slight change in the coefficient of variation of drip system. It can be seen from Table 6.18 that average coefficient of variation after the second application of acid for in-line-type emitter (non-pc) was in the range of 15.10–43.78%. This limit of coefficient of variation was unacceptable for drip irrigation system.

TABLE 6.18   Average Coefficient of Variation before and after the Second Application of Acid

Treatments	CV (%)	
	Before	**After**
	Second Application of Acid	
LT_1	37.22	18.69
LT_2	29.52	15.10
LT_3	31.99	17.35
LT_4	31.58	17.50
LT_5	38.68	18.50
LT_6	43.98	43.78

6.4.3.2.7   AVERAGE OF EMISSION UNIFORMITY BEFORE AND AFTER THE THIRD APPLICATION OF ACID

The data on average EU before and after the third application of acid are presented in Table 6.19 and Fig. 6.24.

TABLE 6.19   Average Emission Uniformity before and after the Third Application of Acid

	EU (%)	
	Before	**After**
Treatments	**Third Application of Acid**	
LT$_1$	71.28	92.60
LT$_2$	80.57	97.84
LT$_3$	73.72	94.90
LT$_4$	76.66	95.55
LT$_5$	70.73	91.81
LT$_6$	38.54	39.78
SE ±	0.711	0.395
CD at 5%	2.238	1.244

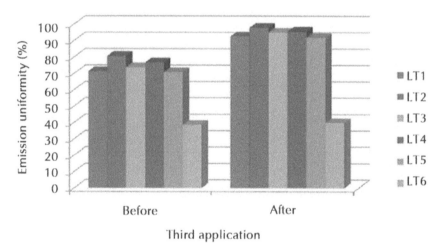

FIGURE 6.24   Emission uniformity (%) for in-line-type emitter before and after the third application of acid.

It can be observed from Table 6.19 that average EU after the third application of acid for in-line-type emitter (non-pc) was in the range of 70.73–80.57%. This limit of EU was marginal for drip irrigation system, due to the dissolution of salts in acids. From Table 6.19, it was concluded that average EU after the third application of acid in LT$_6$ treatment (control) for in-line-type emitter (non-pc) was 38.54%. This limit of EU was unacceptable for drip irrigation system.

After acid and flushing treatments, there was a slight change in the EU of drip system. The highest EU was 97.84% for LT_2 treated with sulfuric acid, thus indicating superiority of sulfuric acid treatment compared with all other treatments. This limit of EU was excellent for drip irrigation system, due to the dissolution of salts in acids. In Table 6.19, average EU after the third application of flushing treatment (control) for in-line-type emitter (non-pc) was 39.78%. This limit of EU is unacceptable for drip irrigation system. These results were significant for drip irrigation system. Similar observations have been reported by other investigators.[3]

6.4.3.2.8 AVERAGE UNIFORMITY COEFFICIENT BEFORE AND AFTER THE THIRD APPLICATION OF ACID

Table 6.20 shows that the average UC after the third application of acid treatments for in-line-type emitter (non-pc) was in the range of 80.96–84.77%. This limit of EU was marginal for drip irrigation system, due to the dissolution of salts in acids. In Table 6.20, average UC after the third application of acid in LT_6 treatment (control) for in-line-type emitter (non-pc) was 51.76%, which was unacceptable for drip irrigation system.

After acid and flushing treatments, there was a slight change in the UC of the drip system. It can be observed from Table 6.20 that the average UC after the third application of acid treatments for in-line-type emitter was in the range of 93.15–97.21%. This limit of UC was excellent for drip irrigation system. Table 6.20 shows average UC after the third application of flushing treatment (control) for in-line-type emitter (non-pc) was 51.45%. This limit of UC was unacceptable for drip irrigation system.

TABLE 6.20 Average Uniformity Coefficient before and after the Third Application of Acid

Treatments	UC (%)	
	Before	**After**
	Third Application of Acid	
LT_1	80.96	93.71
LT_2	84.77	97.29
LT_3	82.11	95.19
LT_4	82.52	95.43
LT_5	81.36	93.15
LT_6	51.76	51.45

6.4.3.2.9   AVERAGE COEFFICIENT OF VARIATION BEFORE AND AFTER THE THIRD APPLICATION OF ACID

It can be observed in Table 6.21 that average coefficient of variation after the third application of acid for in-line-type emitter (non-pc) was in the range of 17.48–48.24%. This limit of coefficient of variation was unacceptable for drip irrigation system.

After acid and flushing treatments, there was a slight change in the coefficient of variation of drip system. Table 6.21 shows that the highest coefficient of variation of 2.71% was found for LT_2 treated with sulfuric acid, thus indicating dominance of sulfuric acid compared with all other treatments. This limit of coefficient of variation was excellent for drip irrigation system. The lower coefficient of variation of 48.55% was found in LT_6 treated with flushing (control). This limit of coefficient of variation was unacceptable for drip irrigation system. Similar observations have been reported by other investigators.[3]

TABLE 6.21   Average Coefficient of Variation before and after the Third Application of Acid

	CV (%)	
	Before	**After**
Treatments	**Third Application of Acid**	
LT_1	19.04	6.29
LT_2	15.23	2.71
LT_3	17.89	4.81
LT_4	17.48	4.57
LT_5	18.64	6.85
LT_6	48.24	48.55

In-line-type non-pressure-compensating (long path) emitters were more susceptible to clogging, due to low discharge of 1.3 lph and turbulent flow regime. These emitters responded well to acid treatments and were reclaimed to the highest level after three applications of acid. Flushing treatment was not found enough for un-clogging.

6.4.4   CROP CHARACTERISTICS

It was found that the emitters were unclogged after the application of acid. Therefore, two poly-houses with substrate cultivation (cocopeat) and soil cultivation were selected to evaluate the effects of unclogging of emitters. Dutch rose flowers were planted in substrate with online type of emitters (pc) and capsicum (bell pepper) was transplanted in soil with in-line type of emitters (non-pc).

6.4.4.1   BIOMETRIC OBSERVATIONS FOR DUTCH ROSE FLOWERS WITH ONLINE-TYPE EMITTERS (PC)

6.4.4.1.1   AVERAGE NUMBER OF SPROUTS

Biometric observations for rose flowers included the number of sprouts, length of stalk, girth of stalk, bud length, and bud diameter during April 16, 2011, to June 25, 2011. The data on the average number of sprouts per plant are presented for each treatment in Table 6.22.

From Table 6.22, it can be observed that maximum number of sprouts was three in OT_2 acid treatment, due to the highest EU that resulted in increased root biomass, and ultimately more absorption of water and nutrients supplied through drip irrigation system.

TABLE 6.22   Average Number of Sprouts per Flower Plant

Treatment	No. of Sprouts
OT_1	2
OT_2	3
OT_3	2
OT_4	2
OT_5	1
OT_6	1
SE ±	0.25
CD at 5%	0.789

However, treatments OT_1, OT_3, and OT_4 were observed with two numbers of sprouts each. In treatment OT_6 (control), minimum number of sprouts was observed, due to low EU in this treatment as compared with all other treatments.

The number of sprouts showed significant differences among all the treatments in this study. The treatment OT_2 was superior with maximum number of sprouts. However, the treatment OT_6 (control) was found with minimum number of sprouts. From these results, it was noticed that the number of sprouts were significantly related to EU of the drip irrigation system.

6.4.4.1.2   AVERAGE STALK LENGTH

The data on average stalk length (cm) of flowers are presented in Table 6.23 and Fig. 6.25.

TABLE 6.23   Average Stalk Length (cm) of Rose Flower

Treatment	\multicolumn{8}{c}{Average Stalk Height (cm)}
	10 DAS	20 DAS	30 DAS	40 DAS	50 DAS	60 DAS	70 DAS	80 DAS
OT_1	4.13	9.06	15.29	25.28	35.53	44.30	53.57	58.86
OT_2	4.96	9.68	16.40	26.97	38.78	48.03	56.57	67.11
OT_3	4.23	9.62	16.59	25.77	37.17	46.10	55.13	64.18
OT_4	4.39	9.76	15.88	26.74	37.93	46.83	55.67	65.61
OT_5	4.02	8.97	15.31	24.79	35.47	44.93	52.61	57.58
OT_6	4.01	8.00	14.57	24.17	30.21	39.01	46.99	54.09
SE ±	4.13	9.06	15.29	25.28	35.53	44.30	53.57	58.86
CD at 5%	0.131	0.226	0.577	0.849	1.306	1.005	1.216	1.72

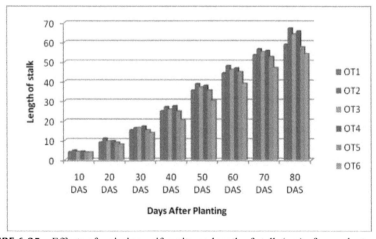

FIGURE 6.25   Effects of emission uniformity on length of stalk (cm) of rose plant.

The average flower stalk length was significantly influenced by different acid treatments. Table 6.23 reveals that the highest stalk length of the plant was found in treatment OT_2 during the observation period. The stalk length was increased from 4.96 cm to 67.11 cm during 80 DAS observation period for OT_2 acid treatment. The OT_2 acid treatment also resulted in highest EU, which resulted in increased root biomass, and ultimately more absorption of water and nutrients and fertilizers supplied through drip irrigation system. The present results are in accordance with Barbosa et al.,[65] who studied the effects of fertigation on quality and yield of rose cv. Sonia and cv. Red success.

The minimum stalk length was 54.09 cm in treatment OT_6 (control) due to uneven distribution of fertilizers and nutrients and at the end of 80 DAS. The grade of flowers was also evaluated by flower stalk length. Flowers with large stalk length are preferred in the market. Among the different treatments, treatment OT_2 was superior among all other treatments.

6.4.4.1.3 AVERAGE STALK GIRTH

The data on average stalk girth (cm) of rose flower are presented in Table 6.24 and in Fig. 6.26.

TABLE 6.24 Average Stalk Girth (cm)

	Average Stalk Girth (cm)							
	Days After Sowing (DAS)							
Treatment	10	20	30	40	50	60	70	80
OT_1	0.2	0.2	0.3	0.3	0.3	0.4	0.5	0.6
OT_2	0.2	0.2	0.3	0.4	0.5	0.6	0.7	0.8
OT_3	0.2	0.2	0.3	0.3	0.3	0.4	0.5	0.6
OT_4	0.2	0.2	0.3	0.4	0.4	0.5	0.6	0.7
OT_5	0.2	0.2	0.3	0.3	0.3	0.4	0.6	0.7
OT_6	0.2	0.2	0.2	0.3	0.3	0.4	0.5	0.6
SE ±	0.021	0.021	0.032	0.029	0.029	0.027	0.029	0.029
CD at 5%	NS	NS	NS	0.091	0.091	0.087	0.091	0.091

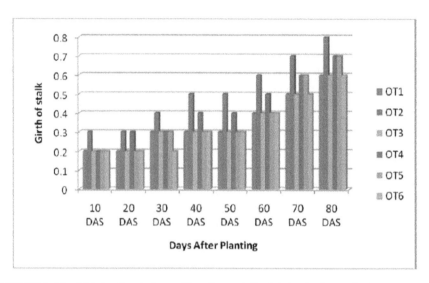

FIGURE 6.26   Effects of emission uniformity on stalk girth (cm) of rose plant.

The average flower stalk girth was significantly influenced by different acid treatments. Table 6.24 reveals that the highest stalk girth was observed in treatment OT_2 during the observation period. The results show that the stalk girth was increased from 0.2 to 0.8 cm during 80 DAS observation period in OT_2 acid treatment. The OT_2 acid treatment resulted in highest EU, which resulted efficient distribution of fertilizers and nutrients, increased root biomass, and ultimately more absorption of water and nutrients supplied through drip irrigation system.

The minimum stalk girth was 0.6 cm in treatments OT_1, OT_3, and OT_5 and at the end of 80 DAS. The grade of flowers was also determined by flower stalk girth. Rose flowers with larger stalk girth are preferred in the market. The present results agree with those by Barbosa.[65] The minimum flower stalk girth was observed in treatment OT_6 (control) due to uneven distribution of fertilizers and nutrients.

6.4.4.1.4   AVERAGE BUD LENGTH AND BUD DIAMETER OF ROSE FLOWER

The data on average bud length (cm) and bud diameter (cm) of flowers are presented in Table 6.25 and Fig. 6.27.

TABLE 6.25 Average Bud Length (cm) and Bud Diameter (cm) of Rose Flower

Treatment	Bud Length of Flower	Bud Diameter
OT_1	3.60	2.35
OT_2	4.07	2.81
OT_3	3.97	2.51
OT_4	4.03	2.72
OT_5	3.53	2.30
OT_6	3.33	2.03
SE ±	0.055	0.036
CD at 5%	0.174	0.115

TABLE 6.26 Average Number of Rose Flowers/m^2

Treatment	No. of Flowers/m^2
OT_1	11.00
OT_2	15.66
OT_3	12.33
OT_4	13.66
OT_5	11.33
OT_6	10.00
SE ±	0.250
CD at 5%	0.788

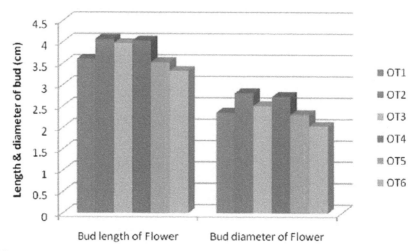

FIGURE 6.27   Effects of emission uniformity on length and diameter of bud (cm) of rose flower.

Table 6.25 shows that the maximum bud length of flower was 4.07 cm in treatment OT_2. The minimum bud length of flower was 3.33 cm in OT_6 (control). The maximum bud diameter of flower was 2.81 cm in treatment OT_2. The minimum bud diameter of flower was 2.03 cm in OT_6 (control) due to uneven distribution of fertilizers and nutrients. The OT_2 acid treatment resulted in highest EU, which resulted in efficient distribution of fertilizers and nutrients, increased roots, biomass, and vegetative growth. Therefore, it resulted in increased bud length and bud diameter of flower.

The statistical analysis on bud length and bud diameter of flower showed significant differences among all the treatment.

6.4.4.1.5   AVERAGE NUMBER OF ROSE FLOWERS

During June 26, 2011, to July 10, 2011, the data on average number of rose flowers/ m^2 are presented in Table 6.26 and Fig. 6.28.

Yield/m2

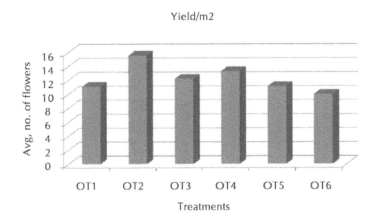

FIGURE 6.28   Effects of emission uniformity on the number of rose flowers (yield/m²).

Table 6.26 shows that maximum number of rose flowers/m² was 15.66 in OT_2. The lowest number of rose flowers/m² was 10.00 in the control treatment. Similar results were reported by Ashok.[63]

Increased EU in OT_2 also resulted in efficient water and nutrient use through drip irrigation thus increasing flower yield. From the observed data, the treatment OT_2 was superior among all the treatments. However, minimum flower yield was obtained in treatment OT_6 control) due to uneven distribution of fertilizers and nutrients.

6.4.4.2   BIOMETRIC OBSERVATIONS FOR CAPSICUM WITH IN-LINE-TYPE EMITTERS (NON-PC)

During April 10, 2011, to June 26, 2011, biometric observations of bell pepper crop included plant height, plant spread, plant girth, length, breadth, and weight of fruit.

6.4.4.2.1   AVERAGE PLANT HEIGHT (CM) OF BELL PEPPER

The data on average plant height (cm) are presented in Table 6.27 and Fig. 6.29. The average plant height was significantly influenced by different acid treatments. Table 6.27 reveals that the highest plant height was observed in treatment LT_2 during the entire observation period. The plant height was increased from 34.53 to 67.40 cm during 90 DAP observation period in LT_2 acid treatment. These results are in accordance with Megharaja.[64] LT_2 acid treatment also resulted in high EU, which resulted in increase in root biomass, and ultimately more absorption of water and nutrients supplied through drip irrigation system.

TABLE 6.27   Average Plant Height (cm) of Bell Pepper Plant

Treatment	Average Plant Height (cm)		
	30 DAP	**60 DAP**	**90 DAP**
LT$_1$	25.73	42.13	54.20
LT$_2$	34.53	53.40	67.40
LT$_3$	30.40	48.47	60.13
LT$_4$	31.53	49.53	63.80
LT$_5$	29.93	47.00	57.80
LT$_6$	17.33	31.33	45.53
SE ±	1.122	1.168	1.606
CD at 5%	4.476	3.676	5.054

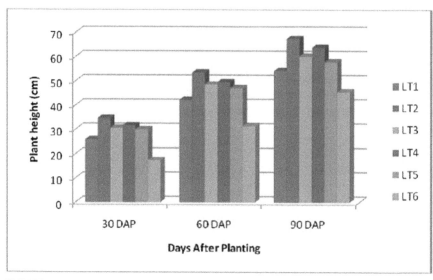

FIGURE 6.29   Effects of emission uniformity on the height (cm) of bell pepper plant.

The minimum plant height was 45.53 cm in treatment LT$_6$ (control) and at the end of 90 DAP. Among all the treatments, LT$_2$ was superior among treatments. The minimum plant height was observed in treatment LT$_6$ (control), due to uneven distribution of fertilizers and nutrients.

6.4.4.2.2 AVERAGE PLANT SPREAD (CM) OF BELL PEPPER

The data on average plant spread in two directions (east–west and north–south) are presented in Table 6.28 and Fig. 6.30.

TABLE 6.28 Average Plant Spread (cm) of Bell Pepper Plant

Treatment	Average Plant Spread (cm)					
	60 DAP		75 DAP		90 DAP	
	EW	NS	EW	NS	EW	NS
LT_1	15.53	17.73	20.60	22.73	28.00	29.07
LT_2	23.67	25.20	29.73	28.00	33.00	35.67
LT_3	20.13	21.67	27.60	25.53	30.33	33.07
LT_4	21.27	22.60	28.07	26.20	31.07	33.73
LT_5	15.00	17.13	19.87	22.27	27.00	28.13
LT_6	14.60	15.40	19.87	20.93	24.40	24.47
SE ±	0.468	0.53	0.547	0.892	0.825	0.531
CD at 5%	1.472	1.669	1.724	2.808	2.598	1.673

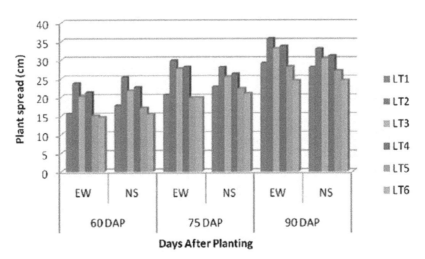

FIGURE 6.30 Effects of emission uniformity of on plant spread (cm) of bell pepper.

Table 6.28 shows that the plant spread in the east–west direction was significantly influenced by different levels of acid treatments, and the highest plant spread in east–west direction was observed in treatment LT_2 during the observation period. The results show that the plant spread in east–west direction was increased from 23.67 to 33.00 cm during 90 DAP in LT_2 acid treatment. The minimum plant spread in east–west direction was 24.4 cm in treatment LT_6 (control) and at the end of 90 DAP, due to uneven distribution of fertilizers and nutrients. Among different treatments, treatment LT_2 was superior, due to increased EU.

Table 6.28 reveals that the highest plant spread in north–south direction was in treatment LT_2 for the entire observation period. The results show that the plant spread in north–south direction was increased from 25.20 to 35.67 cm during 90 DAP observation period in LT_2 acid treatment. The minimum plant spread in north–south direction was 24.47 cm in treatment LT_6 (control) and at the end of 90 DAP. Among different treatments, treatment LT_2 was superior, due to increased EU. The minimum plant spread in north–south direction was in treatment LT_6 (control), due to uneven distribution of fertilizers and nutrients.

6.4.4.2.3 AVERAGE PLANT GIRTH (CM) OF BELL PEPPER

The data on average plant girth (cm) of bell pepper are presented in Table 6.29 and Fig. 6.31.

TABLE 6.29 Average Plant Girth (cm) of Bell Pepper Plant

Treatment	Average Plant Girth (cm)		
	30 DAP	**75 DAP**	**90 DAP**
LT_1	0.8	1.1	1.4
LT_2	1.0	1.3	1.7
LT_3	1.0	1.2	1.5
LT_4	1.0	1.3	1.6
LT_5	0.9	1.0	1.3
LT_6	0.8	0.9	1.0
SE ±	0.038	0.029	0.025
CD at 5%	0.122	0.091	0.078

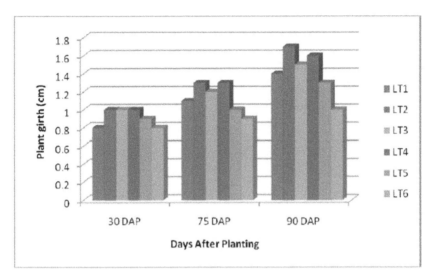

FIGURE 6.31   Effects of emission uniformity on plant girth (cm) of bell pepper.

The average plant girth was significantly influenced by different acid treatments. Table 6.29 indicates that the highest plant girth was observed in treatment LT_2 for the entire observation period. Plant girth was increased from 1.0 to 1.7 cm during 90 DAP observation period in LT_2 acid treatment. The LT_2 acid treatment also resulted in high EU, which resulted in increasing root biomass, and ultimately more absorption of water and nutrients supplied through drip irrigation system.

The minimum plant girth was 1.00 cm in treatment LT_6 (control) and at the end of 90 DAP. Among different treatments, treatment LT_2 was superior. The minimum plant girth was observed in treatment LT_6 (control) due to uneven distribution of fertilizers and nutrients.

6.4.4.2.4   AVERAGE FRUIT LENGTH AND FRUIT BREADTH (CM) OF BELL PEPPER

The data on average fruit length (cm) and fruit breadth (cm) are presented in Table 6.30 and Figs. 6.32 and 6.33.

TABLE 6.30 Average Length, Breadth, and Shape Index (cm) of Bell Pepper Fruit

Treatment	Fruit Length	Fruit Breadth	Shape Index
		cm	
LT_1	7.9	6.2	24.31
LT_2	8.5	6.8	28.93
LT_3	8.1	6.4	25.89
LT_4	8.3	6.5	26.83
LT_5	7.6	6.1	23.22
LT_6	6.5	5.4	17.79
SE ±	0.038	0.037	–
CD at 5%	0.119	0.118	–

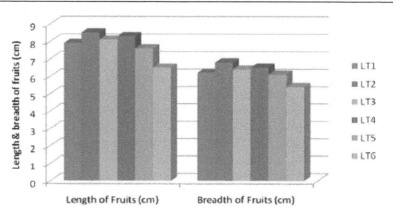

FIGURE 6.32 Effects of emission uniformity on length and breadth (cm) of bell pepper fruit.

FIGURE 6.33 Effects of six acid treatments on length and breadth of bell pepper fruit.

Table 6.30 indicates that the maximum fruit length was 8.5 cm in treatment LT_2. The minimum fruit length was 6.5 cm in LT_6 (control). The maximum fruit breadth was 6.8 cm in treatment LT_2. The minimum fruit breadth was 5.4 cm in LT_6 (control). Similar observations have been reported by Rahman and El-Sheikh.[60]

The statistical analysis on fruit length and fruit breadth showed significant differences among all the treatments.

LT_2 acid treatment resulted in highest EU, which resulted in efficient distribution of fertilizers and nutrients thus leading to increased roots, biomass and vegetative growth, and increased fruit size and weight.

Figure 6.33 shows the effect of six acid treatments on length and breadth of fruit. Treatment LT_2 differs significantly among all treatments.

6.4.4.2.5 AVERAGE SHAPE INDEX OF (CM) BELL PEPPER FRUIT

The data on average shape index (cm) are presented in Table 6.30 and Fig. 6.34.

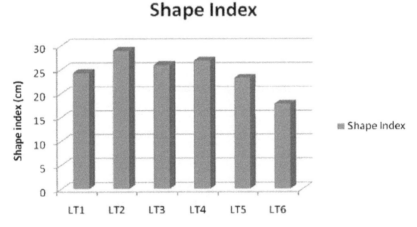

FIGURE 6.34   Effects of emission uniformity on shape index (cm) of bell pepper fruit.

Table 6.30 shows that the maximum shape index of fruit was 28.93 cm in treatment LT_2. The minimum shape index of fruit was 17.79 cm in LT_6 (control), due to uneven distribution of fertilizers and nutrients. The observations are in agreement with those by Rahman and El-Sheikh.[60] LT_2 acid treatment also resulted in highest EU, which resulted in efficient distribution of fertilizers and nutrients thus increasing shape index of fruit. Among different treatments, LT_2 was superior.

6.4.4.2.6   FRUIT YIELD

The data on average yield of fruits/m² (average number of fruit and average fruit weight) are presented in Table 6.31 and Figs. 6.35 and 6.36.

TABLE 6.31   Average Number of Fruits and Weight of Fruits (kg/m²) of Bell Pepper

Treatment	No. of Fruits/m²	Weight kg/m²
LT_1	18.93	2.25
LT_2	25.33	4.25
LT_3	24.53	3.25
LT_4	24.53	3.49
LT_5	18.93	1.96
LT_6	14.40	1.12
SE ±	4.441	0.624
CD at 5%	1.397	1.964

Yield/m2

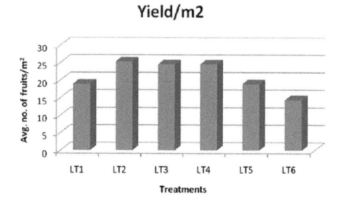

FIGURE 6.35   Effects of emission uniformity on average number of bell pepper fruits/m².

Yield/m2

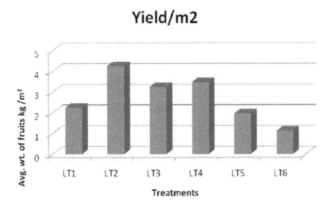

FIGURE 6.36   Effects of emission uniformity on average weight of bell pepper fruits, kg/m^2.

Table 6.31 shows that maximum number of fruits/m^2 was 25.33 in LT$_2$ treatment. Treatment LT$_6$ (control) was found with least number of fruits/m^2 of 14.40. It was observed that maximum weight of fruits (kg/m^2) was 4.25 kg in LT$_2$ treatment. The least weight was 1.12 kg/m^2 in treatment LT$_6$ (control). Similar observations were reported by Buoczlowska.[73]

Increased EU in LT$_2$ treatment resulted in increasing water and nutrient efficiency thereby increasing yield. From the observed data, treatment LT$_2$ was found superior among all the treatments.

6.5   CONCLUSIONS

In India, in majority of the drip-irrigated areas, the source of irrigation water is the groundwater. The main hurdle in drip irrigation management is emitter clogging. Emitter clogging greatly reduces water quality and water distribution uniformity in the irrigated field, which negatively influence crop growth and yield. The emitter clogging can be classified as physical clogging, chemical clogging, and biological clogging. Chemical clogging can be controlled with acid injection by lowering the pH of irrigation water and thus can prevent chemical precipitation. This research study was conducted with the following objectives: (1) To study the efficiency of different acids for the acid treatment of trickle irrigation system; (2) to monitor EU for different acid treatments; (3) to suggest a suitable acid treatment for unclogging of emitters; and (4) to study the crop response under different acid-treated systems. A field experiment on drip irrigation system was conducted in two polyhouses of 20R at Hi-Tech Floriculture Project, Fruit Research Station, Aurangabad, India. Acids under study were DS-99, nitric acid, sulfuric acid, hydrochloric acid, and phosphoric acid. No acid treatment was used as a control. For two poly-houses,

12 treatments (six treatments for online-type emitters and six treatments for in-line-type emitters) were selected. The acid was injected at a fixed interval of 15 days. The biometric observations of plants were plant height (cm) and girth (cm). Major findings of the study were the following:

1. The chemical analysis of water source revealed that the Mg–Ca cationic and Cl–HCO$_3$ anionic were dominant.
2. The pH of water was 7.2, which indicated that the reaction of water was slightly alkaline in nature.
3. For the drip irrigation in this study and at the first acid treatment, EU for online pressure-compensating-type emitters ranged from 72.83 to 74.98% compared with 34.2 to 35.95% for in-line non-compensating-type emitters. In online type of emitter (pc), flow reduction was 26–28% after 7 years of period compared with 65–66% in in-line type of emitter (non-pc) after 6 years of period.
4. At the first acid treatment, UC ranged from 79.9 to 80.84% for online-type emitter (pc) compared with 52.3–56.0% for in-line-type emitter (non-pc).
5. At first acid treatment, the coefficient of variation (C$_V$) ranged from 19.2 to 20.1% for online-type emitter (pc) compared with 43.98–47.7% for in-line-type emitter (non-pc).
6. Online type of emitters unclogged to maximum level after second acid application. The highest EU of 98.11% was found for sulfuric acid (OT$_2$) treatment with the second application for online type of emitter (pc). Due to turbulent flow regime in these emitters, the highest EU was observed after 7 years of continuous use. The highest UC of 98.31% was observed for sulfuric acid (OT$_2$) treatment after the second application for online type of emitter (pc). The highest C$_V$ of 1.69% was observed for sulfuric acid (OT$_2$) treatment after the second application for online type of emitter (pc).
7. In in-line type of emitters (non-pc) with turbulent flow regime, these emitters were more susceptible to clogging. After the third application of acid treatment, the highest EU of 97.84% was observed for sulfuric acid (LT$_2$) for in-line-type emitter (non-pc), indicating dominant effect of sulfuric acid (LT$_2$) treatment compared with all other treatments. There was no considerable change in EU for control treatment. Also UC and C$_V$ for control treatment after the second and third applications were almost unaltered.
8. Statistical analysis indicated that acid treatment was highly significant compared with control (only flushing). Sulfuric acid treatment showed highly significant difference over all other acid treatments for both types of emitters (online type and in-line type).
9. Online-type pressure-compensating emitter (orifice) responded well to acid treatments. Flushing treatment (control) gave less response.
10. In-line-type non-pressure-compensating (long path) emitters were more susceptible to clogging. These emitters responded well to acid treatment

and were reclaimed after three acid applications to the highest level. Flushing treatment (control) was not found suitable.

11. Sulfuric acid was found most effective for unclogging of emitters followed by hydrochloric acid, nitric acid, DS-99 acid, and phosphoric acid. Emitters with low-discharge non-pc and drip system in last phase were required to be treated at 15 days of interval for three times, where emitters with high discharge and pc drip system needed only two acid applications. Flushing application was not suitable for drip irrigation system.

12. The values of stalk length, stalk girth, bud length, and bud diameter of rose were higher in sulfuric acid (OT_2) treatment than other treatments, in online-type emitters (pc).

13. The values of plant height, plant girth, plant spread, fruit length, and fruit breadth of capsicum plants were higher in sulfuric acid (LT_2) treatment than other treatments in in-line-type emitters (non-pc).

14. Effect of six treatments on fruit length and fruit breadth in sulfuric acid (LT_2) differed significantly than all other treatments. It was also found that EU of drip irrigation was different among all acid treatments. Control (LT_6) treatment gave low quality of fruits due to the low EU.

15. The highest yield of rose (number of flowers/m²) was 15.66/m² was for sulfuric acid (OT_2) for online-type emitter (pc), thus showing dominant effect of sulfuric acid (OT_2) treatment, compared with all other treatments.

16. The highest yield of capsicum (weight of fruits/m²) was 4.25 kg/m² for sulfuric acid (LT_2) for in-line-type emitter (non-pc), thus showing dominant effect of sulfuric acid (LT_2) treatment compared with all other treatments.

6.6 SUMMARY

The main hurdle in drip irrigation system management is emitter clogging. Emitter clogging greatly reduces the water distribution uniformity in the irrigated field, which negatively influences crop growth and yield. Considering the problem of clogging in drip irrigation system, a field experiment "studies on efficiency of acids for unclogging of emitters" was conducted in two poly-houses of 20R each at Hi-Tech floriculture project, Fruit Research Station, Aurangabad.

Acids for the study were DS-99, nitric acid, sulfuric acid, hydrochloric acid, and phosphoric acid. No acid treatment was taken as a control. For selected two poly-houses, total 12 treatments (six treatments for online-type emitter and six treatments for in-line-type emitter) were considered. Further, it was also planned to inject the acid at fixed interval (15 days). In one poly-house, pot culture in coco-peat (substrate) cultivation (Dutch Roses) with Netafim online pressure-compensating drip irrigation system was installed. In the other poly-house with soil cultivation, Red soil mix (capsicum) with Jain in-line non-pressure-compensating drip irrigation system was used. The common fertilizer doses were applied for both crops. The biometric observations of plants were plant height (cm) and girth (cm) to see the effects of unclogging emitters due to application of different acid treatments.

Online-type pressure-compensating emitters (orifice, pc; 8 lph with turbulent flow) were less susceptible to clogging although the drip irrigation system was 7 years old. In-line-type non-pressure-compensating (long path, non-pc; 1.3 lph with turbulent flow regime) emitters were more susceptible to clogging. Sulfuric acid was most effective for unclogging of emitters followed by hydrochloric acid, nitric acid, DS-99 acid, and phosphoric acid, respectively. Emitter with low-discharge non-pc and drip system in the last phase needed 15 days of interval for three times of treatment, whereas emitter with high discharge and pc drip system needed only two acid applications, and flushing application was not suitable for unclogging of emitters.

The highest yield of rose (number of flowers/m^2) was 15.66/m^2 for sulfuric acid (OT$_2$) for online-type emitter (pc), which shows the dominant effect of sulfuric acid (OT$_2$), compared with all other treatments. The highest yield of capsicum (weight of fruits/m^2) was 4.25 kg/m^2 was for sulfuric acid (LT$_2$) for in-line-type emitter (non-pc), which shows the effect of sulfuric acid (LT$_2$) treatment, compared with all other treatments.

KEYWORDS

- Acidification
- ASAE
- Bell pepper
- Capsicum
- clogging
- coefficient of variation
- drip irrigation
- dripper
- emission uniformity
- emitter
- field evaluation
- saline water
- India
- Maharashtra
- micro irrigation
- rose flower
- uniformity coefficient

REFERENCES

1. Anonymous. International floriculture trade. *J. Environ. Manage.* 2006, 76, 238–341.
2. Kong, Y. *Effect of Operating Pressure on Micro Irrigation Uniformity*; Yucheng Comprehensive Experimental Station, Institute of Geographical Science and Natural Resources Research, Chinese Academy of Sciences: Beijing, P. R. China, 2000.
3. Anonymous. *Performance of Drip Irrigation Systems Under Field Conditions*; ARC-Institute for Agricultural Engineering: Silverton, South Africa, 1997.
4. Von Zobeltitz. Greenhouse farming. *Am. Soc. Plast.*, 1999, 64, 131–136.
5. Mundada, P. R. Assessment of Quality of Well Water in Kopergaon Tahsil of Ahmednagar District. M. Sc. Thesis, Mahatma Phule Krishi Vidyapeeth, Rahuri, 1990.
6. Bargues, J. P.; Barragan, J. Effect of flushing frequency on emitter clogging in micro irrigation with effluents. *Agric. Water Manage.* 2005, 97, 883–891.
7. Capra, A.; Tamburino, F. Water quality and distribution uniformity in drip irrigation system. *J. Agric. Eng. Res.* 1995, 70, 355–365.
8. Taylor, G. *Injection Rates and Components of a Fertigation System*; College of Tropical Agriculture and Human Resources, University of Hawaii, Manoa, USA, Bulletin EN-4, 1995.
9. Zang, F. The role of physical – chemical treatment in wastewater reclamation and reuse. The Hebrew University of Jerusalem, Jerusalem 91904, Israel. *Water Sci. Tech.* 2007, 37(10), 79–90.
10. Bucks, D. A. Principles, practices and potentialities of trickle irrigation. *Appl. Eng. Agric. (ASAE)*, 1979, 1, 219–225.
11. Wu, I. P. Drip irrigation uniformity considering emitter plugging. *Trans. ASAE* 2004, 24(5), 1234–1240.
12. Nakayama, F. S.; Bucks, D. Emitter clogging effect on trickle irrigation uniformity. *Trans. ASAE* 1991, 24(1), 77–80.
13. Dehghanisanji, A. The effect of chlorine on emitter clogging by algae and protozoa and the performance of drip irrigation. *Trans. ASAE* 2004, 48(2), 519–527.
14. Ravina, K. Control of emitter clogging in drip irrigation with reclaimed wastewater. *Irrig. Sci.* 1992, 13, 129–139.
15. Anonymous. *Micro Irrigation*; Jain Irrigation Systems Ltd, 2002.
16. Argus. *Argus Nutrient Dosing Handbook*; Argus Control System Ltd.: White Rock, BC, Canada, 2009.
17. Obreza, T. A. *Maintenance Guide for Florida Micro Irrigation Systems*; University of Florida, IFAS Cooperative Extension Service, Circular 1449, 2004.
18. Oktem, A. Effect of different irrigation intervals to drip irrigated dent corn (Zea mays L. indentata): water – yield relationship. *Pak. J. Biol. Sci.* 2006, 9(8), 1476–1481.
19. Ustun, S.; Anapali, O.; Donmez, M. F. Biological treatment of clogged emitters in a drip irrigation system. *J. Environ. Manage.* 2005, 76, 238–341.
20. Evans, R.; Probsting, K. *Micro Irrigation*; Washington State University, Irrigated Agriculture Research and Extension Center: Prosser, WA, USA, 1990.
21. Pampattiwar, P. S. *Progress and Perspective of Drip Irrigation*; Symposia on Drip Irrigation at Mahatma Phule Agricultural University, Rahuri, Publication 54, 1994, p 12.
22. Modi, S. Water quality in drip irrigation system. *J. Agric. Eng. Res.* 2001, 70, 355–365.
23. William, W. *Irrigation Efficiency and Uniformity and Crop Water Use Efficiency*; The Board of Regents of the University of Nebraska, 2002.
24. Narayanmoorthy, A. *Potential for Drip and Sprinkler Irrigation System*; Gokhale Institute of Politics and Economics (Deemed University), Pune-411004, 2005.
25. Oza, A. *Irrigation and Water Resources*; India Infrastructure Research, 2007.

26. Awulachew, S. B.; Talu, T. *Irrigation Methods: Drip Irrigation Options for Smallholders*; International Water Management Institute, Module 5 Part II, 2009.

27. SAI Platform, 2010. *Water Conservation Technical Briefs: TB 8 – Use of Drip Irrigation*. SAI Platform.

28. David, M.; Mills, L. *Drip Irrigation System*; University of Nevada Cooperative Extension Service, 1992.

29. Toth, A. *Drip Irrigation*; Queen Gil Tape, 2004.

30. Rogers, D. H.; Lamm, F. R.; Alam, M. *Subsurface Drip Irrigation System Water Quality Assessment Guideline*; Kansas State University, 2003.

31. Arizona Landscape Irrigation Guidelines Committee. *Guidelines for Landscape Drip Irrigation Systems*; AMWUA Organization, 2006.

32. Cox, D. M. *Drip Irrigation System*; Cooperative Agric. Extension at University of Nevada, Fact Sheet 91-53, 1995.

33. Mota, R. *Introduction to Ambiental Engineering*; ABES: Rio de Janeiro, 1997; p 292.

34. Zoldoske, D. F. *Drip and Micro Irrigation*; California State University: Fresno, USA, 2009.

35. Patil, R. D. Quality of well water as influenced by location along the canal & season. *J. Soils Crops* 1993, 3(2), 128–131.

36. Clark, G. A.; Rogers, G. *Maintaining Drip Irrigation Systems*; Kansas State University Agricultural Experiment Station and Cooperative Extension Service, Circular 96-440-E, 1996.

37. Schultheis, B. *Maintenance of Drip Irrigation Systems*; University of Missouri Extension: Marshfield, MO, 1999.

38. Lamont, W. J.; Orzolek, M. D. *Drip Irrigation for Vegetable Production*; PSU, College of Agricultural Sciences, Agricultural Research and Cooperative Extension, US Department of Agriculture, 2002.

39. Lamont, W. J. *Maintaining Drip Irrigation Systems*; Department of Horticulture, College of Agric. Sciences, Pennsylvania State University, 2005.

40. Thokal, R. T.; Mahale, D. M. *Drip Irrigation System: Clogging and Prevention*; Pointer Publisher: Rajasthan, India, 2004; pp 32–40.

41. Shrivastava, R. Water quality in drip irrigation system. *J. Agric. Eng. Res.* 2006, 70, 125–128.

42. Flynn, R. *Irrigation Water Analysis and Interpretation*; US Department of Agriculture Cooperating, New Mexico State University, User's Guide W-102, 2009.

43. Rowan, M. A. *The Utility of Drip Irrigation for the Distribution of On-Site Waste Water Effluent*; Department of Food, Agricultural and Biology Engineering, The Ohio State University, 2004.

44. Runyan, C. *Maintenance Guide for Micro Irrigation Systems in the Southern Region*; New Mexico State University, 2002.

45. Lamm, F. R. Field Evaluation of Micro Irrigation Systems. In Paper Presentation at the Section Meeting of ASBE; KSU Northwest Research Extension Center: Colby, Kansas, USA, 1997.

46. Soccol, O. J.; Ullmann, M. N.; Frizzone, J. A. Performance analysis of a trickle irrigation subunit installed in an apple orchard. *Int. J. Brazilian Archives Biol. Technol.* 2002, 45(4), 525–530.

47. Reinders, F. B. *Performance of Drip Irrigation Systems Under Field Conditions*; Operational Program, ARC Institute for Agricultural Engineering: Silverton, South Africa, 2005.

48. El Gendy, R. W.; Gadalla, A. M.; Hamdy, A. *Irrigation Water Saving Via Scheduling Irrigation of Snap Bean and Direction of Soil Water Movement Under Drip Irrigation System*; Soil and Water Res. Dept., Nuclear Res. Center, AEA: Cairo, Egypt, 2007.

49. Bakhsh, A.; Kandhawa, U. A.; Ishaug, W. Deficit irrigation effects on cotton yield using drip irrigation system. *Pak. J. Water Resour.* 2008, 12(2).

50. Aali, K. A.; Liaghat, A. The effect of acidification and magnetic field on emitter clogging under saline water application. *J. Agric. Sci.* 2009, 1(1).
51. Sah, D. H.; Purohit, R. C.; Jain, S. K. Design, construction and evaluation of low pressure and low cost drip irrigation system. *Int. Agric. Eng. J.* 2010, 19(2).
52. Alam, M.; Rogers, D. H. *Filtration and Maintenance Considerations for Subsurface Drip Irrigation Systems*; Kansas State University, Agricultural Experiment Station and Cooperative Extension Service: Manhattan, Kansas, 2002.
53. Bozkurt, S.; Ozekici, B. The effect of fertigation management on clogging of in-line emitters. University of Mustafa Kemal, 31780 Samandag, Hatay, Turkey. *J. Appl. Sci.* 2006, 6(15), 3026–3034.
54. Ribeiro, T. A. P.; Paterniani, J. E. S.; Coletti, C. Chemical treatment to unclog dripper irrigation systems due to biological problems. *Sci. Agric. (Piracicaba, Braz.)* 2008, 65(1), 1–9.
55. Qingsong, W.; Shuhuai, H. Evaluations of emitter clogging in drip irrigation by two-phase flow simulations and laboratory experiments. *Comput. Electron. Agric.*, 2008, 63, 294–303.
56. Liu, H.; Huang, G. *Laboratory Experiment on Drip Emitter Clogging with Fresh Water and Treated Sewage Effluent*; Agricultural Water Management: Beijing, China, 2008.
57. Yavuz, M. Y.; Erken, O.; Bahar, E. Emitter clogging and effects on drip irrigation system performance. *Afr. J. Agric. Res.* 2010, 5(7), 532–538.
58. Enciso, J.; Porter, D. *Maintaining Subsurface Drip Irrigation System*; Texas Cooperative Extension at the Texas A & M University Systems, Leaflet 5401:10-01, 2001.
59. Netafim, *Drip System Operation and Maintenance*; Netafim Irrigation: USA, 2009. www.netafimusa.com.
60. Rahman, A. S. Z.; El-Sheikh, T. M. A comparative study on some sweet pepper cultivation grown under plastic house conditions for yield and storability. *Egypt. J. Hortic.* 1994, 21(2), 213–225.
61. Seekar, I.; Hochmuth, G. Low tunnels for early watermelon production in North Florida. *Am. Soc. Plasticulture*, 1994, 64, 131–136.
62. Saen, K.; Pathom, N. *Effect of Pruning on Yield and Quality of Sweet Pepper*; 1999. www.acr.avrdc.Org.
63. Ashok, A. D. Influence of graded levels and sources of N fertigation on flowering of cut rose cv. first red under protected conditions. *South Indian Horticulture*, 1999, 47(1–6), 115–118.
64. Megharaja, K. M. Studies on the Effect of Growing Conditions and Growth Regulations on Growth & Productivity of Hybrid Capsicum cv. Indira. M.Sc. (Agri.) Thesis, Univ. Agric. Sci., Bangalore, Karnataka, India, 2000.
65. Barbosa, J. G.; Santose, J. A. Quality and Commercial Grade of Rose Yield as Affected by Potassium Application Through Drip Irrigation, ISHS Acta Horticulture, 4th International Symposium on Rose Research and Cultivation, 2005, p 751.
66. Zubair, M.; Zafar, M. Effect of potassium on pre-flowering growth of gladiolus cultivars. *J. Agric. Biol. Sci.* 2006, 1(3), 36–46.
67. Bralts, V. F. Drip irrigation design and evaluation based on statistical uniformity concept. *Adv. Irrig.* 1983, 4(2), 67–117.
68. ASAE. *Design and Installation of Micro Irrigation Systems*; EP 405.1, ASAE Standards, 2002; pp 903–907.
69. ASAE. *EP458: Field Evaluation of Micro Irrigation System*; ASAE Standards, 1997.
70. Richard, G. *Drip Irrigation – An Overview*; 2006. www.prairie – element.ca.
71. English, S. D. Filtration and water treatment for micro irrigation. In *Drip Irrigation in Action*; ASAE: St. Joseph, MI, USA, 1985; Vol. 1; pp 50–57.
72. Darbie, M. *Drip Chemigation: Injecting Fertilizers, Acid & Chlorine*; The University of Georgia, College of Agricultural & Environment Sciences and the U.S. Department of Agriculture, 2005.

73. Buoczkowska, H. Evaluation of yield of six sweet pepper cultivators grown in an unheated foil tunnel and in the open field. *Folia Hortic.* 1990, 2, 29–39.
74. Sachin, U. Biological treatment of clogged emitters in a drip irrigation system. *J. Environ. Manage.* 2005, 76, 338–341.

APPENDICES

Modified and reprinted with permission from: Goyal, M. R., 2012. Appendices. Pp. 317–332. In: *Management of Drip/Trickle or Micro Irrigation,* edited by M. R. Goyal. NJ, USA: Apple Academic Press Inc.

APPENDIX A

Conversion of SI and Non-SI Units

To Convert Column 1 to Column 2, Multiply by	Column 1, SI Unit	Column 2, Non-SI Unit	To Convert Column 2 to Column 1, Multiply by
Linear			
0.621	kilometer, km (10^3 m)	miles, mi	1.609
1.094	meter, m	yard, yd	0.914
3.28	meter, m	feet, ft	0.304
3.94×10^{-2}	millimeter, mm (10^{-3})	inch, in	25.4
Squares			
2.47	hectare, he	acre	0.405
2.47	square kilometer, km^2	acre	4.05×10^{-3}
0.386	square kilometer, km^2	square mile, mi^2	2.590
2.47×10^{-4}	square meter, m^2	acre	4.05×10^{-3}
10.76	square meter, m^2	square feet, ft^2	9.29×10^{-2}
1.55×10^{-3}	mm^2	square inch, in^2	645

Cubics

9.73×10^{-3}	cubic meter, m^3	inch-acre	102.8
35.3	cubic meter, m^3	cubic-feet, ft^3	2.83×10^{-2}
6.10×10^4	cubic meter, m^3	cubic inch, in^3	1.64×10^{-5}
2.84×10^{-2}	liter, L (10^{-3} m^3)	bushel, bu	35.24
1.057	liter, L	liquid quarts, qt	0.946
3.53×10^{-2}	liter, L	cubic feet, ft^3	28.3
0.265	liter, L	Gallon, –	3.78
33.78	liter, L	fluid ounce, oz	2.96×10^{-2}
2.11	liter, L	fluid dot, dt	0.473

Weight

2.20×10^{-3}	gram, g (10^{-3} kg)	pound,	454
3.52×10^{-2}	gram, g (10^{-3} kg)	ounce, oz	28.4
2.205	kilogram, kg	pound, lb	0.454
10^{-2}	kilogram, kg	quintal (metric), q	100
1.10×10^{-3}	kilogram, kg	ton (2000 lbs), ton	907
1.102	megagram, mg	ton (US), ton	0.907
1.102	metric ton, t	ton (US), ton	0.907

Yield and rate

0.893	kilogram per hectare	pound per acre	1.12

7.77×10^{-2}	kilogram per cubic meter	pound per fanega	12.87
1.49×10^{-2}	kilogram per hectare	pound per acre, 60 lb	67.19
1.59×10^{-2}	kilogram per hectare	pound per acre, 56 lb	62.71
1.86×10^{-2}	kilogram per hectare	pound per acre, 48 lb	53.75
0.107	liter per hectare	gallon per acre	9.35
893	ton per hectare	pound per acre	1.12×10^{-3}
893	megagram per hectare	pound per acre	1.12×10^{-3}
0.446	ton per hectare	ton (2000 lb) per acre	2.24
2.24	meter per second	mile per hour	0.447

Specific surface

| 10 | square meter per kilogram | square centimeter per gram | 0.1 |
| 10^3 | square meter per kilogram | square millimeter per gram | 10^{-3} |

Pressure

9.90	megapascal, MPa	atmosphere	0.101
10	megapascal	bar	0.1
1.0	megagram per cubic meter	gram per cubic centimeter	1.00
2.09×10^{-2}	pascal, Pa	pound per square feet	47.9

1.45×10^{-4}	pascal, Pa	pound per square inch	6.90×10^{3}

Temperature

$1.00 (K - 273)$	Kelvin, K	centigrade, °C	$1.00 (C+273)$
$(1.8 C + 32)$	centigrade, °C	Fahrenheit, °F	$(F - 32)/1.8$

Energy

9.52×10^{-4}	Joule J	BTU	1.05×10^{3}
0.239	Joule, J	calories, cal	4.19
0.735	Joule, J	feet-pound	1.36
2.387×10^{5}	Joule per square meter	calories per square centimeter	4.19×10^{4}
10^{5}	Newton, N	dynes	10^{-5}

Water requirements

9.73×10^{-3}	cubic meter	inch acre	102.8
9.81×10^{-3}	cubic meter per hour	cubic feet per second	101.9
4.40	cubic meter per hour	gallon (US) per minute	0.227
8.11	hectare-meter	acre-feet	0.123
97.28	hectare-meter	acre-inch	1.03×10^{-2}
8.1×10^{-2}	hectare centi-meter	acre-feet	12.33

Concentration

1	centimole per kilogram	milliequiva-lents per	
	100 grams	1	
0.1	gram per kilo-gram	percents	10
1	milligram per kilogram	parts per mil-lion	1

Nutrients for plants			
2.29	P	P_2O_5	0.437
1.20	K	K_2O	0.830
1.39	Ca	CaO	0.715
1.66	Mg	MgO	0.602

NUTRIENT EQUIVALENTS

		Conversion	Equivalent
Column A	Column B	A to B	B to A
	NH3	1.216	0.822
	NO3	4.429	0.226
	KNO3	7.221	0.1385
	Ca(NO3)2	5.861	0.171
	(NH4)2SO4	4.721	0.212
	NH4NO3	5.718	0.175
N	(NH4)2 HPO4	4.718	0.212
	P2O5	2.292	0.436
	PO4	3.066	0.326
	KH2PO4	4.394	0.228
	(NH4)2 HPO4	4.255	0.235
P	H3PO4	3.164	0.316
	K2O	1.205	0.83
	KNO3	2.586	0.387
	KH2PO4	3.481	0.287
	KCl	1.907	0.524
K	K2SO4	2.229	0.449

	CaO	1.399	0.715
	Ca(NO3)2	4.094	0.244
	CaCl2 × 6H2O	5.467	0.183
Ca	CaSO4 × 2H2O	4.296	0.233
	MgO	1.658	0.603
Mg	MgSO4 × 7H2O	1.014	0.0986
	H2SO4	3.059	0.327
	(NH4)2 SO4	4.124	0.2425
	K2SO4	5.437	0.184
	MgSO4 × 7H2O	7.689	0.13
S	CaSO4 × 2H2O	5.371	0.186

APPENDIX B

Pipe and Conduit Flow

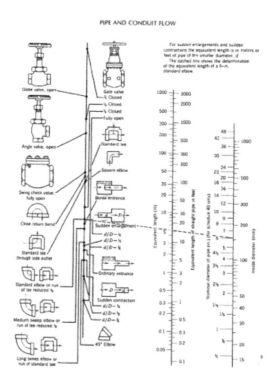

PIPE AND CONDUIT FLOW

APPENDIX C

Percentage of Daily Sunshine Hours: for North and South Hemispheres

Latitude	Jan	Feb	Mar	Apr	May	Jun	Jul	Aug	Sep	Oct	Nov	Dec
North												
0	8.50	7.66	8.49	8.21	8.50	8.22	8.50	8.49	8.21	8.50	8.22	8.50
5	8.32	7.57	8.47	3.29	8.65	8.41	8.67	8.60	8.23	8.42	8.07	8.30
10	8.13	7.47	8.45	8.37	8.81	8.60	8.86	8.71	8.25	8.34	7.91	8.10
15	7.94	7.36	8.43	8.44	8.98	8.80	9.05	8.83	8.28	8.20	7.75	7.88
20	7.74	7.25	8.41	8.52	9.15	9.00	9.25	8.96	8.30	8.18	7.58	7.66
25	7.53	7.14	8.39	8.61	9.33	9.23	9.45	9.09	8.32	8.09	7.40	7.52
30	7.30	7.03	8.38	8.71	9.53	9.49	9.67	9.22	8.33	7.99	7.19	7.15
32	7.20	6.97	8.37	8.76	9.62	9.59	9.77	9.27	8.34	7.95	7.11	7.05
34	7.10	6.91	8.36	8.80	9.72	9.70	9.88	9.33	8.36	7.90	7.02	6.92
36	6.99	6.85	8.35	8.85	9.82	9.82	9.99	9.40	8.37	7.85	6.92	6.79
38	6.87	6.79	8.34	8.90	9.92	9.95	10.1	9.47	3.38	7.80	6.82	6.66
40	6.76	6.72	8.33	8.95	10.0	10.1	10.2	9.54	8.39	7.75	6.72	7.52
42	6.63	6.65	8.31	9.00	10.1	10.2	10.4	9.62	8.40	7.69	6.62	6.37
44	6.49	6.58	8.30	9.06	10.3	10.4	10.5	9.70	8.41	7.63	6.49	6.21
46	6.34	6.50	8.29	9.12	10.4	10.5	10.6	9.79	8.42	7.57	6.36	6.04
48	6.17	6.41	8.27	9.18	10.5	10.7	10.8	9.89	8.44	7.51	6.23	5.86
50	5.98	6.30	8.24	9.24	10.7	10.9	11.0	10.0	8.35	7.45	6.10	5.64
52	5.77	6.19	8.21	9.29	10.9	11.1	11.2	10.1	8.49	7.39	5.93	5.43
54	5.55	6.08	8.18	9.36	11.0	11.4	11.4	10.3	8.51	7.20	5.74	5.18
56	5.30	5.95	8.15	9.45	11.2	11.7	11.6	10.4	8.53	7.21	5.54	4.89
58	5.01	5.81	8.12	9.55	11.5	12.0	12.0	10.6	8.55	7.10	4.31	4.56
60	4.67	5.65	8.08	9.65	11.7	12.4	12.3	10.7	8.57	6.98	5.04	4.22
South												
0	8.50	7.66	8.49	8.21	8.50	8.22	8.50	8.49	8.21	8.50	8.22	8.50
5	8.68	7.76	8.51	8.15	8.34	8.05	8.33	8.38	8.19	8.56	8.37	8.68
10	8.86	7.87	8.53	8.09	8.18	7.86	8.14	8.27	8.17	8.62	8.53	8.88
15	9.05	7.98	8.55	8.02	8.02	7.65	7.95	8.15	8.15	8.68	8.70	9.10
20	9.24	8.09	8.57	7.94	7.85	7.43	7.76	8.03	8.13	8.76	8.87	9.33
25	9.46	8.21	8.60	7.74	7.66	7.20	7.54	7.90	8.11	8.86	9.04	9.58
30	9.70	8.33	8.62	7.73	7.45	6.96	7.31	7.76	8.07	8.97	9.24	9.85
32	9.81	8.39	8.63	7.69	7.36	6.85	7.21	7.70	8.06	9.01	9.33	9.96

34	9.92	8.45	8.64	7.64	7.27	6.74	7.10	7.63	8.05	9.06	9.42	10.1
36	10.0	8.51	8.65	7.59	7.18	6.62	6.99	7.56	8.04	9.11	9.35	10.2
38	10.2	8.57	8.66	7.54	7.08	6.50	6.87	7.49	8.03	9.16	9.61	10.3
40	10.3	8.63	8.67	7.49	6.97	6.37	6.76	7.41	8.02	9.21	9.71	10.5
42	10.4	8.70	8.68	7.44	6.85	6.23	6.64	7.33	8.01	9.26	9.8	10.6
44	10.5	8.78	8.69	7.38	6.73	6.08	6.51	7.25	7.99	9.31	9.94	10.8
46	10.7	8.86	8.90	7.32	6.61	5.92	6.37	7.16	7.96	9.37	10.1	11.0

APPENDIX D

Psychrometric Constant (γ) for Different Altitudes (Z)

$$\gamma = 10^{-3}\,[(C_p P)/(\varepsilon \lambda)] = (0.00163) \times [P/\lambda]$$

where γ is the Psychrometric constant (kPa C^{-1}); C_p is the specific heat of moist air = 1.013 (kJ kg^{-10}C^{-1}); P is the atmospheric pressure (kPa); ε is the ratio molecular weight of water vapor/dry air = 0.622; λ is the latent heat of vaporization = 2.45 MJ kg^{-1} at 20°C.

Z m	γ kPa/°C	Z m	γ kPa/°C	Z m	γ kPa/°C	Z m	γ kPa/°C
0	0.067	1000	0.060	2000	0.053	3000	0.047
100	0.067	1100	0.059	2100	0.052	3100	0.046
200	0.066	1200	0.058	2200	0.052	3200	0.046
300	0.065	1300	0.058	2300	0.051	3300	0.045
400	0.064	1400	0.057	2400	0.051	3400	0.045
500	0.064	1500	0.056	2500	0.050	3500	0.044
600	0.063	1600	0.056	2600	0.049	3600	0.043
700	0.062	1700	0.055	2700	0.049	3700	0.043
800	0.061	1800	0.054	2800	0.048	3800	0.042
900	0.061	1900	0.054	2900	0.047	3900	0.042
1000	0.060	2000	0.053	3000	0.047	4000	0.041

APPENDIX E

Saturation Vapor Pressure [e_s] for Different Temperatures (T)

Vapor pressure function = $e_s = [0.6108] \times \exp\{[17.27 \times T]/[T + 237.3]\}$

T °C	e_s kPa	T °C	e_s kPa	T °C	e_s kPa	T °C	e_s kPa
1.0	0.657	13.0	1.498	25.0	3.168	37.0	6.275
1.5	0.681	13.5	1.547	25.5	3.263	37.5	6.448
2.0	0.706	14.0	1.599	26.0	3.361	38.0	6.625
2.5	0.731	14.5	1.651	26.5	3.462	38.5	6.806
3.0	0.758	15.0	1.705	27.0	3.565	39.0	6.991
3.5	0.785	15.5	1.761	27.5	3.671	39.5	7.181
4.0	0.813	16.0	1.818	28.0	3.780	40.0	7.376
4.5	0.842	16.5	1.877	28.5	3.891	40.5	7.574
5.0	0.872	17.0	1.938	29.0	4.006	41.0	7.778
5.5	0.903	17.5	2.000	29.5	4.123	41.5	7.986
6.0	0.935	18.0	2.064	30.0	4.243	42.0	8.199
6.5	0.968	18.5	2.130	30.5	4.366	42.5	8.417
7.0	1.002	19.0	2.197	31.0	4.493	43.0	8.640
7.5	1.037	19.5	2.267	31.5	4.622	43.5	8.867
8.0	1.073	20.0	2.338	32.0	4.755	44.0	9.101
8.5	1.110	20.5	2.412	32.5	4.891	44.5	9.339
9.0	1.148	21.0	2.487	33.0	5.030	45.0	9.582
9.5	1.187	21.5	2.564	33.5	5.173	45.5	9.832
10.0	1.228	22.0	2.644	34.0	5.319	46.0	10.086
10.5	1.270	22.5	2.726	34.5	5.469	46.5	10.347
11.0	1.313	23.0	2.809	35.0	5.623	47.0	10.613
11.5	1.357	23.5	2.896	35.5	5.780	47.5	10.885
12.0	1.403	24.0	2.984	36.0	5.941	48.0	11.163
12.5	1.449	24.5	3.075	36.5	6.106	48.5	11.447

APPENDIX F

Slope of Vapor Pressure Curve (Δ) for Different Temperatures (T)

$\Delta = [4098.\ e^0(T)]/[T + 237.3]^2$							
$= 2504\{\exp[(17.27T)/(T + 237.2)]\}/[T + 237.3]^2$							
T °C	Δ kPa/°C	T °C	Δ kPa/°C	T °C	Δ kPa/°C	T °C	Δ kPa/°C
1.0	0.047	13.0	0.098	25.0	0.189	37.0	0.342
1.5	0.049	13.5	0.101	25.5	0.194	37.5	0.350
2.0	0.050	14.0	0.104	26.0	0.199	38.0	0.358
2.5	0.052	14.5	0.107	26.5	0.204	38.5	0.367
3.0	0.054	15.0	0.110	27.0	0.209	39.0	0.375
3.5	0.055	15.5	0.113	27.5	0.215	39.5	0.384
4.0	0.057	16.0	0.116	28.0	0.220	40.0	0.393
4.5	0.059	16.5	0.119	28.5	0.226	40.5	0.402
5.0	0.061	17.0	0.123	29.0	0.231	41.0	0.412
5.5	0.063	17.5	0.126	29.5	0.237	41.5	0.421
6.0	0.065	18.0	0.130	30.0	0.243	42.0	0.431
6.5	0.067	18.5	0.133	30.5	0.249	42.5	0.441
7.0	0.069	19.0	0.137	31.0	0.256	43.0	0.451
7.5	0.071	19.5	0.141	31.5	0.262	43.5	0.461
8.0	0.073	20.0	0.145	32.0	0.269	44.0	0.471
8.5	0.075	20.5	0.149	32.5	0.275	44.5	0.482
9.0	0.078	21.0	0.153	33.0	0.282	45.0	0.493
9.5	0.080	21.5	0.157	33.5	0.289	45.5	0.504
10.0	0.082	22.0	0.161	34.0	0.296	46.0	0.515
10.5	0.085	22.5	0.165	34.5	0.303	46.5	0.526
11.0	0.087	23.0	0.170	35.0	0.311	47.0	0.538
11.5	0.090	23.5	0.174	35.5	0.318	47.5	0.550
12.0	0.092	24.0	0.179	36.0	0.326	48.0	0.562
12.5	0.095	24.5	0.184	36.5	0.334	48.5	0.574

APPENDIX G

Number of the Day in the Year (Julian Day)

Day	Jan	Feb	Mar	Apr	May	Jun	Jul	Aug	Sep	Oct	Nov	Dec
1	1	32	60	91	121	152	182	213	244	274	305	335
2	2	33	61	92	122	153	183	214	245	275	306	336
3	3	34	62	93	123	154	184	215	246	276	307	337
4	4	35	63	94	124	155	185	216	247	277	308	338
5	5	36	64	95	125	156	186	217	248	278	309	339
6	6	37	65	96	126	157	187	218	249	279	310	340
7	7	38	66	97	127	158	188	219	250	280	311	341
8	8	39	67	98	128	159	189	220	251	281	312	342
9	9	40	68	99	129	160	190	221	252	282	313	343
10	10	41	69	100	130	161	191	222	253	283	314	344
11	11	42	70	101	131	162	192	223	254	284	315	345
12	12	43	71	102	132	163	193	224	255	285	316	346
13	13	44	72	103	133	164	194	225	256	286	317	347
14	14	45	73	104	134	165	195	226	257	287	318	348
15	15	46	74	105	135	166	196	227	258	288	319	349
16	16	47	75	106	136	167	197	228	259	289	320	350
17	17	48	76	107	137	168	198	229	260	290	321	351
18	18	49	77	108	138	169	199	230	261	291	322	352
19	19	50	78	109	139	170	200	231	262	292	323	353
20	20	51	79	110	140	171	201	232	263	293	324	354
21	21	52	80	111	141	172	202	233	264	294	325	355
22	22	53	81	112	142	173	203	234	265	295	326	356
23	23	54	82	113	143	174	204	235	266	296	327	357
24	24	55	83	114	144	175	205	236	267	297	328	358
25	25	56	84	115	145	176	206	237	268	298	329	359
26	26	57	85	116	146	177	207	238	269	299	330	360
27	27	58	86	117	147	178	208	239	270	300	331	361
28	28	59	87	118	148	179	209	240	271	301	332	362

29	29	(60)	88	119	149	180	210	241	272	302	333	363
30	30	–	89	120	150	181	211	242	273	303	334	364
31	31	–	90	–	151	–	212	243	–	304	–	365

APPENDIX H

Stefan–Boltzmann Law at Different Temperatures (*T*)

$[\sigma^*(T_K)^4] = [4.903 \times 10^{-9}]$, MJ K^{-4} m^{-2} day^{-1}

where $T_K = \{T[°C] + 273.16\}$

T °C	$\sigma^*(T_K)^4$ MJ m^{-2} d^{-1}	T °C	$\sigma^*(T_K)^4$ MJ m^{-2} d^{-1}	T °C	$\sigma^*(T_K)^4$ MJ m^{-2} d^{-1}
1.0	27.70	17.0	34.75	33.0	43.08
1.5	27.90	17.5	34.99	33.5	43.36
2.0	28.11	18.0	35.24	34.0	43.64
2.5	28.31	18.5	35.48	34.5	43.93
3.0	28.52	19.0	35.72	35.0	44.21
3.5	28.72	19.5	35.97	35.5	44.50
4.0	28.93	20.0	36.21	36.0	44.79
4.5	29.14	20.5	36.46	36.5	45.08
5.0	29.35	21.0	36.71	37.0	45.37
5.5	29.56	21.5	36.96	37.5	45.67
6.0	29.78	22.0	37.21	38.0	45.96
6.5	29.99	22.5	37.47	38.5	46.26
7.0	30.21	23.0	37.72	39.0	46.56
7.5	30.42	23.5	37.98	39.5	46.85
8.0	30.64	24.0	38.23	40.0	47.15
8.5	30.86	24.5	38.49	40.5	47.46
9.0	31.08	25.0	38.75	41.0	47.76
9.5	31.30	25.5	39.01	41.5	48.06
10.0	31.52	26.0	39.27	42.0	48.37
10.5	31.74	26.5	39.53	42.5	48.68

11.0	31.97	27.0	39.80	43.0	48.99
11.5	32.19	27.5	40.06	43.5	49.30
12.0	32.42	28.0	40.33	44.0	49.61
12.5	32.65	28.5	40.60	44.5	49.92
13.0	32.88	29.0	40.87	45.0	50.24
13.5	33.11	29.5	41.14	45.5	50.56
14.0	33.34	30.0	41.41	46.0	50.87
14.5	33.57	30.5	41.69	46.5	51.19
15.0	33.81	31.0	41.96	47.0	51.51
15.5	34.04	31.5	42.24	47.5	51.84
16.0	34.28	32.0	42.52	48.0	52.16
16.5	34,52	32.5	42.80	48.5	52.49

APPENDIX I

Thermodynamic Properties of Air and Water

1. Latent Heat of Vaporization (λ)
$$\lambda = [2.501 - (2.361 \times 10^{-3})\ T]$$
where λ is the latent heat of vaporization (MJ kg^{-1}) and T is the air temperature (°C).

The value of the latent heat varies only slightly over normal temperature ranges. A single value may be taken (for ambient temperature = 20°C): $\lambda = 2.45$ MJ kg^{-1}.

2. Atmospheric Pressure (P)
$$P = P_0\ [\{T_{K0} - \alpha(Z - Z_0)\ \} / \{T_{K0}\}]^{(g/(\alpha.R))}$$
where P is the atmospheric pressure at elevation Z (kPa); P_0 is the atmospheric pressure at sea level = 101.3 kPa; Z, elevation (m); Z_0 is the elevation at reference level (m); g, gravitational acceleration = 9.807 m s^{-2}; R is the specific gas constant = 287 J kg^{-1} K^{-1}; α is the constant lapse rate for moist air = 0.0065 K m^{-1}; T_{K0} si the reference temperature (K) at elevation Z_0 = 273.16 + T; T is the air temperature for the time period of calculation (°C).

When assuming P_0 = 101.3 kPa at Z_0 = 0, and T_{K0} = 293 K for T = 20°C, the above equation reduces to
$$P = 101.3[(293 - 0.0065Z)\ (293)]^{5.26}$$

3. Atmospheric Density (ρ)
$$\rho = [1000P]/[T_{Kv}\ R] = [3.486P]/[T_{Kv}],\ \text{and}\ T_{Kv} = T_K[1 - 0.378(e_a)/P]^{-1}$$

where ρ is the atmospheric density (kg m^{-3}); R, specific gas constant = 287 J kg^{-1}K^{-1}; T_{Kv}, virtual temperature (K); T_K absolute temperature (K), $T_K = 273.16 + T$ (°C); e_a, actual vapor pressure (kPa); T, mean daily temperature for 24-h calculation time steps.

For average conditions (e_a in the range of 1–5 kPa and P between 80 and 100 kPa), T_{Kv} can be substituted by $T_{Kv} \approx 1.01$ (T + 273)

4. Saturation Vapor Pressure Function (e_s)
$$e_s = [0.6108] \times \exp\{[17.27 \times T]/[T + 237.3]\}$$

where e_s is the saturation vapor pressure function (kPa) and T is the air temperature (°C)

5. Slope Vapor Pressure Curve (Δ)
$$\Delta = [4098. \ e^0(T)]/[T + 237.3]^2$$
$$= 2504\{\exp[(17.27T)/(T + 237.2)]\}/[T + 237.3]^2$$

where Δ is the slope vapor pressure curve (kPa C^{-1}); T is the air temperature (°C); and e^0(T) is the saturation vapor pressure at temperature T (kPa).

In 24-h calculations, Δ is calculated using mean daily air temperature. In hourly calculations, T refers to the hourly mean, T_{hr}.

6. Psychrometric Constant (γ)
$$\gamma = 10^{-3} [(C_p.P)/(\varepsilon.\lambda)] = (0.00163) \times [P/\lambda]$$

where γ is the psychrometric constant (kPa C^{-1}); C_p is the specific heat of moist air = 1.013 kJ kg^{-10}C^{-1}; P, atmospheric pressure (kPa): eq 2 or 4; ε is the ratio molecular weight of water vapor/dry air = 0.622; and λ is the latent heat of vaporization (MJ kg^{-1})

7. Dew Point Temperature (T_{dew})
When data is not available, T_{dew} can be computed from e_a by
$$T_{dew} = [\{116.91 + 237.3 \text{Log}_e(e_a)\}/\{16.78 - \text{Log}_e(e_a)\}]$$

where T_{dew} is the dew point temperature (°C); and e_a, actual vapor pressure (kPa)
For the case of measurements with the Assmann psychrometer, T_{dew} can be calculated from
$$T_{dew} = (112 + 0.9T_{wet})[e_a/(e^0 \ T_{wet})]^{0.125} - [112 - 0.1T_{wet}]$$

8. Shortwave Radiation on a Clear Sky Day (R_{so})

The calculation of R_{so} is required for computing net long-wave radiation and for checking the calibration of pyranometers and the integrity of R_{so} data. A good approximation for R_{so} for daily and hourly periods is

$R_{so} = (0.75 + 2 \times 10^{-5} Z)R_a$

where Z is the station elevation (m); and R_a is the extraterrestrial radiation (MJ m^{-2} d^{-1}).

Equation is valid for station elevations less than 6000 m having low air turbidity. The equation was developed by linearizing Beer's radiation extinction law as a function of station elevation and assuming that the average angle of the sun above the horizon is about 50°.

For areas of high turbidity caused by pollution or airborne dust or for regions where the sun angle is significantly less than 50° so that the path length of radiation through the atmosphere is increased, an adoption of Beer's law can be employed where P is used to represent atmospheric mass:

$R_{so} = (R_a) \exp[(-0.0018P)/(K_t \sin(\Phi))]$

where K_t is the turbidity coefficient, $0 < K_t \le 1.0$ where $K_t = 1.0$ for clean air and $K_t = 1.0$ for extremely turbid, dusty, or polluted air; P, atmospheric pressure (kPa); Φ, angle of the sun above the horizon (rad); R_a is the extraterrestrial radiation (MJ m^{-2} d^{-1})

For hourly or shorter periods, Φ is calculated as

$\sin \Phi = \sin \varphi \sin \delta + \cos \varphi \cos \delta \cos \omega$

where φ is the latitude (rad); δ is the solar declination (rad) (eq. 24 in Chapter 3); and ω is the solar time angle at midpoint of hourly or shorter period (rad).

For 24-h periods, the mean daily sun angle, weighted according to R_a, can be approximated as

$\sin(\Phi_{24}) = \sin[0.85 + 0.3 \varphi \sin\{(2\pi J/365)-1.39\}-0.42 \varphi^2]$

where Φ_{24} is the average Φ during the daylight period, weighted according to R_a (rad); φ, latitude (rad); and J, day in the year.

The Φ_{24} variable is used to represent the average sun angle during daylight hours and has been weighted to represent integrated 24-h transmission effects on 24-h R_{so} by the atmosphere. Φ_{24} should be limited to ≥ 0. In some situations, the estimation for R_{so} can be improved by modification to consider the effects of water vapor on shortwave absorption, so that $R_{so} = (K_B + K_D)R_a$ where

$K_B = 0.98\exp[\{(-0.00146P)/(K_t \sin \Phi)\}-0.091\{w/\sin \Phi\}^{0.25}]$

where K_B is the clearness index for direct beam radiation and K_D is the corresponding index for diffuse beam radiation.

$K_D = 0.35-0.33 K_B$ for $K_B \ge 0.15$
$K_D = 0.18 + 0.82 K_B$ for $K_B < 0.15$

R_a, extraterrestrial radiation (MJ m^{-2} d^{-1}); K_t, turbidity coefficient, $0 < K_t \le 1.0$ where $K_t = 1.0$ for clean air and $K_t = 1.0$ for extremely turbid, dusty, or polluted air; P, atmospheric pressure (kPa); Φ, angle of the sun above the horizon (rad); W, perceptible water in the atmosphere (mm) = 0.14 e_a P + 2.1; e_a, actual vapor pressure (kPa); P, atmospheric pressure (kPa).

APPENDIX J

Psychrometric Chart at Sea Level

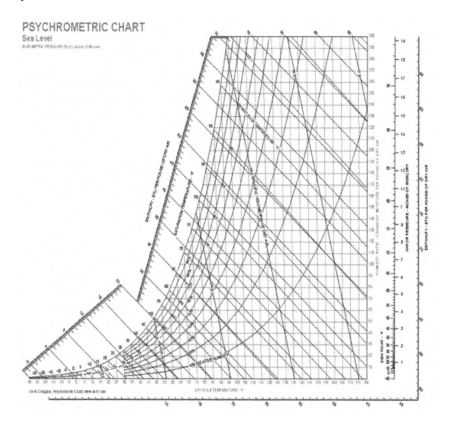

APPENDIX K

[http://www.fao.org/docrep/T0551E/t0551e07.htm#5.5%20field%20management%20practices%20in%20wastewater%20irrigation]

1. Relationship between applied water salinity and soil water salinity at different leaching fractions (FAO, 1985).

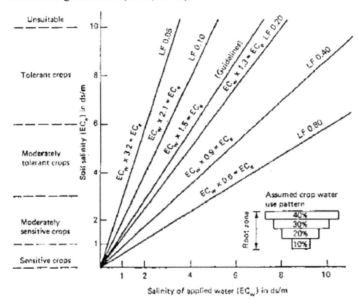

2. Schematic representations of salt accumulation, planting positions, ridge shapes, and watering patterns.

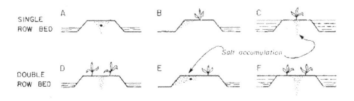

3. Main components of general planning guidelines for wastewater reuse (Cobham and Johnson, 1988).

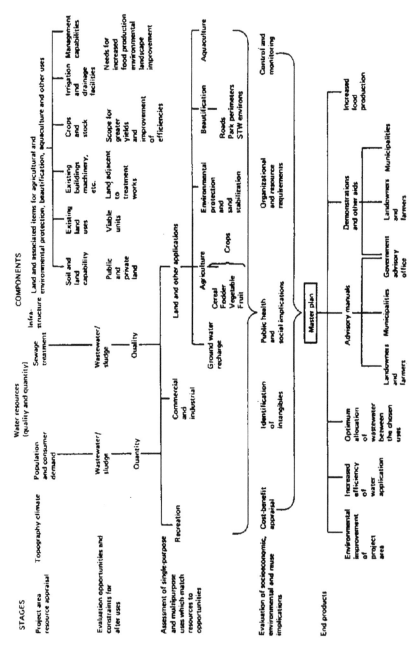

APPENDIX L

From: *Vincent F. Bralts*, 2015. Chapter 3: Evaluation of the uniformity coefficients.

In: *Sustainable Micro Irrigation Management for Trees and Vines, Volume 3* by M. R. Goyal (Ed.). Apple Academic Press Inc.,

1. Uniformity Classification

Classification	Statistical Uniformity	Emission Uniformity
Excellent	For U = 100–95%	100–94%
Good	For U = 90–85%	87–81%
Fair	For U = 80–75%	75–68%
Poor	For U = 70–65%	62–56%
Not acceptable	For U < 60%	<50%

2. Acceptable Intervals of Uniformity in a Drip Irrigation System

Type of Dripper	Slope	Uniformity Interval, %
Point source: located in planting distance > 3.9 m	Level[a]	90–95
	Inclined[b]	85–90
Point source: located in planting distance < 3.9 m	Level[a]	85–90
	Inclined[b]	80–90
Drippers inserted in the lines for annual row crops	Level[a]	80–90
	Inclined[b]	75–85

[a]Level = slope less that 2%.
[b]Inclined = slope greater than 2%

3. Confidence Limits for Field Uniformity (U)

Field Uniformity (%)	18 Drippers		36 Drippers		72 Drippers	
	Confidence Limit		Confidence Limit		Confidence Limit	
	N Sum[a]	%	N Sum	%	N Sum	%
100	3	$U \pm 0.0$	6	$U \pm 0.6\%$	12	$U \pm 0.0\%$
90	3	$U \pm 2.9$	6	$U \pm 2.0\%$	12	$U \pm 1.4\%$
80	3	$U \pm 5.8$	6	$U \pm 4.0\%$	12	$U \pm 2.8\%$
70	3	$U \pm 9.4$	6	$U \pm 6.5\%$	12	$U \pm 4.5\%$
60	3	$U \pm 13.3$	6	$U \pm 9.2\%$	12	$U \pm 6.5\%$

[a]**N Sum** = One-sixth part of the total measured drippers. This is a number of samples that will be added to calculate T_{max} and T_{min}.

4. Nomograph for statistical uniformity.

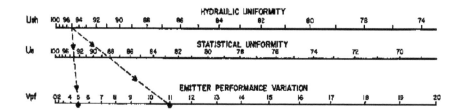

5. The field uniformity of an irrigation system based on the dripper times and the dripper flow rate.

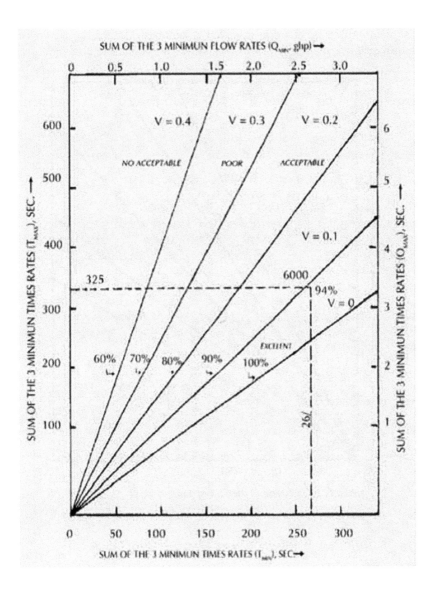

6. The field uniformity of a drip irrigation system based on the time to collect a known quantity of water or based on pressure for hydraulic uniformity.

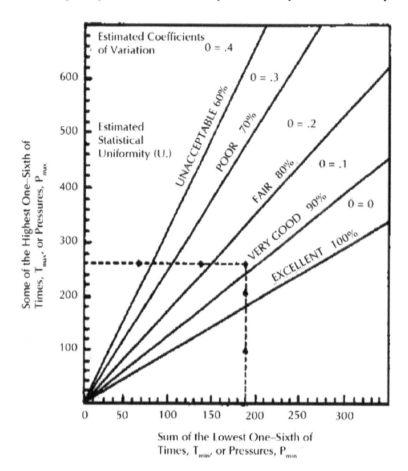

INDEX

Milton Keynes UK
Ingram Content Group UK Ltd.
UKHW031145141024
449569UK00024B/1060